To Get The Lights

A Memoir about

Farm Electrification in Saskatchewan

Dave Anderson

For Marguita & James
With my best wishes
Dave Anderson.

To Get The Lights

Also by David John Anderson, 1925 - The Seventh Daughter - a family biography - self published in 1995.
Saskatchewan Archives Board -
Accession No. R95-52 - File R-E3954.

Cover photographs by the author.

Note for Librarians: A cataloguing record for this book is available from Library and Archives Canada at www.collectionscanada.ca/amicus/index-e.html
ISBN 1-4120-6371-x

green power

Printed in Victoria, BC, Canada. Printed on paper with minimum 30% recycled fibre. Trafford's print shop runs on "green energy" from solar, wind and other environmentally-friendly power sources.

TRAFFORD
PUBLISHING
Offices in Canada, USA, Ireland and UK
This book was published *on-demand* in cooperation with Trafford Publishing. On-demand publishing is a unique process and service of making a book available for retail sale to the public taking advantage of on-demand manufacturing and Internet marketing. On-demand publishing includes promotions, retail sales, manufacturing, order fulfilment, accounting and collecting royalties on behalf of the author.

Book sales for North America and international:
Trafford Publishing, 6E–2333 Government St.,
Victoria, BC v8t 4p4 CANADA
phone 250 383 6864 (toll-free 1 888 232 4444)
fax 250 383 6804; email to orders@trafford.com
Book sales in Europe:
Trafford Publishing (uk) Ltd., Enterprise House, Wistaston Road Business Centre,
Wistaston Road, Crewe, Cheshire cw2 7rp UNITED KINGDOM
phone 01270 251 396 (local rate 0845 230 9601)
facsimile 01270 254 983; orders.uk@trafford.com
Order online at:
trafford.com/05-1282

10 9 8 7 6 5 4 3 2

This book is dedicated in loving memory to my Mother and Father; to Betty, who has supported this venture from beginning to end; and with love to my six children, David, Patricia, Mark, Guy, Nancy and Lisa.

". . . Two roads diverged in a yellow wood And I – I took the one less travelled by, And that has made all the difference."

Robert Frost

CONTENTS

Preface

HISTORIANS HAVE CHRONICLED that electrification of rural Saskatchewan happened in the 1950s and '60s by the CCF government's crown corporation, the Saskatchewan Power Corporation (SPC.) However this achievement is usually glossed over in a few sentences, maybe a paragraph or two, with a dearth of detail about how the job was done. Succeeding generations of sod-busters take for granted the presence of electricity across length and breadth of the Wheat Province. This accomplishment has been described by some in the power utility business as a miracle. But miracles do not just happen. It is my hope that *To Get The Lights* will give some insights into how the huge obstacles of distances and finances were overcome with the significant, volunteer cooperative effort of hundreds in the farming community to achieve this miracle.

Sharon Butala, a well-known, respected and published Saskatchewan author of *Lilac Moon* and *The Perfection Of The Morning*, after reading a snippet of this work wrote, "This is a very interesting piece of important Saskatchewan history." Her comment was a huge encouragement for me and for that I am grateful.

Writing *To Get The Lights* has been, for me, a true labour of love on two counts. First, I loved my farm electrification work and, second, penning these pages has been hard labour. My journalistic skills are suspect because if writers needed licences I'd have trouble getting one. Nor, at 80, can I say age diminishes my writing ability. I thank Andrea Scott-Bigsby for copy-editing these pages and for her many suggestions for their improvement, proof that I are no grammarian.

I started to write this composite of adventures in 1995 as I moved from the prime of life towards old age. What started out as satisfying a curiosity evolved into a demanding mistress, then a passion leading to complete absorption and, finally, an obsession. They are just one darn story after another and, as well, they meander, but these are my stories and meandering is just part of me. As these pages filled they took on a psychoanalytical role, disturbing on some days, but I have enjoyed reliving the episodes that touched me and were such a big part of my life. Those years could die and be interred with what's left of me but it seems sad to depart this life without my loved ones having some understanding of my labours during this brief stay. These threads of my life's tapestry may fill voids in the customary expressionless dash between date of birth and death which will be engraved somewhere, some day.

To help readers understand me, I offer the following insights.

* * *

IN 1925, WHEN I WAS BORN, my parents lived in Tribune, a tiny, prairie hamlet well off the beaten path in south-eastern Saskatchewan, near the United States border. My arrival into this world was in hospital in the nearby city of Estevan. In 1928 we moved a few miles further west to Assiniboia. My first remembrances as an only child start there amid the Great Depression. Assiniboia was at the centre of Saskatchewan's "dustbowl," unquestionably the most severely ravaged part of Canada during the Dirty Thirties. They were disastrous times when the prairie economy, already devastated by the country's financial woes, was savaged even further by nature in unyielding droughts which fried crops year after year. When green shoots did show, they were devoured by hordes of ravenous insects and rodents.

Those calamitous times indelibly wounded my parents, as happened to so many other self-reliant, hard-working, decent people. One of their desperation ventures was to run a rooming and boarding house rented from Clint Smith, a New York Ranger hockey player. This endeavour was abandoned in 1933. It cost more to operate than the roomers and boarders could pay – four school teachers, who lived with us, and several others, up to eleven at times, who ate there. Amongst the boarders were two RCMP officers charged to rid the town of the ever-present hoboes. Our house was either known in the encampments of those who rode the rails or secretly marked in the back lane as a food haven. Mother's compassion for these hungry derelicts was great. She was unable to spurn their need and fed them at the back door, sometimes as the constables filled their bellies at our dining room table. An indelible example for an eight-year-old that laws are not always just.

Another social justice lesson I learned came in 1934, when I was nine. It is proof that one person can make a difference. The town's paddling pool was left empty that summer to save water costs. With more imagination than toys one afternoon, some pals and I trapped a few gophers for their tails, which paid us a penny each from the rural municipality. Later, we skinny-dipped in the Canadian Pacific Railway dugout that supplied water to their steam engines in a nearby roundhouse. We did this without our parents knowing, and it was dangerous, as most, including me, were non-swimmers. Accidentally, one of my friends, seven-year-old Leo Lesperance slipped into deep water and drowned. The next day, July 7, likely at Mother's nudge, I started a petition to "Town Counsle" to reverse their pool decision. Signed by 27 kids I presented it to the Town Clerk. To my delight the pool was filled but, sadly, it was too late for Leo, his grieving mother, sister and two brothers.

Worse than the tough times and our poverty was Dad's devastating loss of hope, unable to find work of any kind. A jack-of-all-trades, he had several jobs until the businesses closed or went broke. His unbridled optimism was fueled with the hope that President Franklin D. Roosevelt's "New Deal" for the USA, his birthplace, would spill over into Canada. But that failed to happen. Those dispensing government "relief" denied him sustenance, felt he could do so many different jobs as compared to the singular skills of the town's accountants, lawyers, dentists and doctors, already on public assistance. Compounding this bizarre qualification was the shame he must have felt when, as breadwinner, he could no longer provide for his wife and son and was forced to give up, desperate and exhausted by the hopelessness of his deprivation. Politicians at every level simply didn't care, particularly those in Ottawa, who had little or no comprehension of the dire conditions in the West. That my parents endured against these odds for as long as they did is a tribute to their tenacity and determination, but their limits had been breached by the Depression when all their options vanished.

While shielded from the abject misery which saddened their lives, I never felt hard-done-by. I knew nothing different, took our poverty for granted. My friends were all in the same boat. I witnessed this when the McBoyle family invited me to their home for dinner. It consisted of a half cup of dry, uncooked oatmeal, a bit of sugar and some milk. As I grew up those hard times were not discussed, nor did I query my parents. An exception, a telling detail I inadvertently overheard Mother admit to her Nova Scotia family, was that in the toughest times she watered down my milk to make it go further.

In 1936 those cruel times required me to live for two and a half years with my devout, maternal grandparents, in their 80s, and my three spinster aunts, who would never turn away one of their own kind. The Dirty Thirties did not spare them, but they were better off, able to grow and barter for much of their basic needs. Theirs was not a high living standard but a high life standard in Bear River, a rural Nova Scotia village in the Annapolis Valley, near the often ornery Bay of Fundy, where they had raised a family of 13. The daily care of an active and curious grandson must have presented huge challenges they neither needed nor wanted. My adjustments were significant but theirs must have been monumental. Mother kept house and toiled for a florist across the "Bay" in Saint John for her board and room and five dollars a week, half of which went to her parents for my keep. Dad vainly tried to sell washing machines and vacuum cleaners in Regina. Both hoped someday we three could reunite.

My nearly blind grandfather, unquestioned family patriarch and disciplinarian, became my father figure at the impressionable age of 11. Known locally as "The Captain," he had sailed the oceans for 50 years until retirement at 75. He told me about getting his sea legs on sailing vessels, and taught me about the thrust of the tides, of storms and spellbinding tales of the sea. Regrettably, none were recorded, now only vaguely remembered, and are lost forever. This shipwreck of family history has made writing these pages important for me.

His high moral standards and inflexible expectations influenced me greatly. Many of my values are the result of his nurture, whether I liked it or not. He was grateful I became his eyes. We were together a lot and talked: at mealtimes, while walking to and from church and the barber shop, while chopping and piling wood, and while working in the shed, barn, orchard and garden. He unwittingly overlooked my age, treated me like an adult. Our companionship was special. My chores, an hour or so daily, helped offset my keep. A lesson learned was when you worked you worked. Blisters and calluses were not strangers to me. Another was to conserve, not just for a rainy day, but in everything all of the time, a trait that has not diminished within me over time.

While my Nova Scotia days were mostly happy times I missed my parents and friends. In spite of our close proximity, visits with my Mother were infrequent. I wasn't homesick because we had no home, but I often wrestled with loneliness, a feeling akin to a disease. So there were some sad days, and nights, in my straw-ticked bed in an upstairs hallway of a household with such a strict sense of frugality, order and discipline. In spite of their unquestioned love for me, and me for them, it wasn't the same as being with my parents. I longed for them, felt forsaken and bewildered that they were not there to guide, reassure and comfort me.

Those Bear River days significantly affected my later behaviour as a husband and father, molded my sense of identity, made me fiercely independent and resourceful, stimulated an active imagination and nature as a very private person. We reunited with Dad in Regina just before Christmas in 1938 but the once strong father-son relationship never recovered from our estrangement.

After Mother's death in 1979 and as Dad's health waned from several strokes, I found several hundred of her letters around which I wove her biography, *The Seventh Daughter*. Putting those pages together helped me to better understand the magnitude of their ordeal and the subtle impact the Great Depression had on me. Their graves hold many unanswered questions, a void which has influenced this venture. I am grateful they gave me invaluable inheritances of curiosity, honesty, work ethics and unflinching optimism. Growing up in those days gave me a considerable head of steam, the likes of which are foreign to many young people today.

*   *   *

AMONGST MY PERSONAL TRAITS are those of initiative and determination that were doubly ingrained in me in Regina in the spring of 1939, as the agonies imposed by the Great Depression started to dissipate and the government mysteriously found money and material to fight a war. At 13 and in high school, I wanted to earn some spending money.

The first instance arose from a Foodland advertisement, "Needed: a boy with bicycle and carrier." The pay: five cents per grocery box delivered anywhere in the city. I lined up with more than 60 others that Saturday morning. As each entered the owner's office a couple of questions were asked, then the lad was ushered out, followed by the next. When my turn came, I noticed a short wooden ruler on the floor so picked it up and placed it on his desk. The errant ruler was his hiring criteria. He told the rest to leave. Then, to me, he said, "You saw something that needed doing, and you did it. You're the boy I need." I think those for whom I later worked saw me in a similar light.

The second instance happened the same year. I wanted a newspaper route. It would mean regular and more income. The Leader Post said that to get one, I had to sell new subscriptions. So on May first I jumped on my bike and pedalled after moving vans. When they stopped to unload I sales-pitched the new occupant. By nightfall many had turned me down but I had signed three new customers. Bus Derbyshire, Circulation Manager, said, "If you can sell three new starts in this day and age, you deserve a route." My assignment was route 130, 37 papers on McIntyre Street's west side, from Victoria Avenue to College Avenue. It earned me about two dollars a week. I could then pay for my clothes and school supplies, open a bank account and contribute a dollar a month to my parents toward household expenses.

* * *

THE ROYAL CANADIAN NAVY stole part of my youth in the Second World War and, along with some unkind seas and awful war zone experiences, quickly matured me before it ended. I was only 19 when it came to a close. Despite the dehumanizing environment of wartime military training I was exposed to the rigidity of obedience and discipline, plus a sense of duty to my country and ship-mates, and of honour to my pals already in uniform, several of whom paid the ultimate price. As a signalman on the ship's bridge alongside captains and officers, instilled in me were their leadership qualities and commands that were not the basis for any discussion or argument. Those experiences significantly affected my personality and behaviour.

* * *

I HAD A REWARDING 35-year career in Canada's energy industry. When snippets of my work experiences were told to family or friends, a comment was often made that I should write a book. So the notion is not new.

To Get The Lights covers the first seven years of my work with SPC. Those long days in the 1950s, when farm electrification was in full swing and brought electricity to Saskatchewan farmers for the first time, were the fondest and most rewarding of my entire career. It was a time when society was less imbued with ownership whims of consumer goods and the excesses of today's commercialism with its throwaway mentality. A time when people were less self-centered and skeptical, were kinder, more respectful, trustful and tolerant of one another. A time that was more genteel – life was simpler and folks were more dependent on their neighbours than on government. A time that was less competitive than today's rude, unforgiving and selfish era. A time when we were able to do so much so quickly to improve the quality of life for so many, a task that in today's environment would not have been possible.

They were also times of unrelenting growth for SPC because of the ever-increasing use of electricity by existing consumers, combined with the incessant demands to add new customers in hundreds of new communities and on tens of thousands of farms as power lines fanned out across the province. Amidst these challenges of growth the utility also had to operate, to keep the lights on. But system operations are much more routine, even mundane, as compared to times of rapid expansion and the innovations needed to accommodate the unprecedented growth.

SaskPower today gives only oblique recognition to those pioneers who overcame the adversities which brought the Corporation's early accomplishments to fruition in the first place. But then, present day staff don't know their company's history. Most don't care. They want to reinvent the wheel, and are more imbued with what lies ahead than what lessons the past may teach.

Mine were not pivotal roles nor was I part of major company decisions that shaped its course. While I have no illusions of the importance of my achievements I did put a lot of me into SPC's accomplishments. I will not grace any history page about the Wheat Province's energy story. My career is not the stuff of triumphant management annals. I was not blessed with profound wisdom nor the gift of strategic thought process. Others with whom I worked had better minds, were bigger thinkers. My career, during this time of rapid expansion, spanned an adventurous period of giant technological leaps in data processing and communications, of change and innovation, the pressure of tough targets and of erring then trying again.

When I joined SPC, a technocratic juggernaut, I knew the path I chose was not strewn with roses and that generous doses of ingenuity and hard work would be needed for a successful career. An axiom imprinted on me by my grandfather, "Anything worth doing is worth doing well," was recognized by my superiors. Ted Durnin, SPC's Construction Engineer, once said to others in my presence, "Give it to Dave, He'll get it done." This trait earned a reputation on which my career thrived.

My ferocious work ethic entombed and defined me more than was reasonable. It adversely affected my first marriage. My wife, at times justifiably disdainful of my employer, as many spouses are of demanding bosses, resented and lost all interest in my work so its successes and failures were not shared at home as my children grew up. More time should have been spent with them but it seemed more important to be in Porcupine Plain or Pemmican Portage or Prairie River. I selfishly chased my career. Any self esteem I had came from work. What follows tells of a big part of me, but my private life details are omitted.

*   *   *

MY JUNE BIRTHDAY June birthday labels me a Gemini, characterized as having a twin personality, a bewildering dichotomy of one, a private, solitary, independent person and the other, an outspoken, assertive, ambitious and enthusiastic extrovert. With a passion and visceral pleasure for hard work, as I was taught, I had a bone-deep belief to work as diligent for my employer as I would have for myself. My fierce independence is fuelled by persistence and determination, to the point of stubbornness, never more so than when a principle or decency was at stake. I am an optimist with an uncontrollable weakness to see the good side of an unpromising situation. And to top it all off, I had more than my share of luck and good health.

New and unfamiliar roles never daunted me, nor did making decisions. Always ready for a new challenge it seems I followed Yogi Berra's wisdom, "When you come to a fork in the road, take it." Some new spheres were stressful but each had its own challenges and the decisions necessary to assuage them. I grabbed at every opportunity, well aware that the success pyramid was a steep and slippery slope.

My freewheeling nature likely made me a difficult subordinate. I am impetuous, verbal and outspoken with a sense of bravado, not afraid to speak bluntly or reluctant to challenge the status quo or avoid controversy. My confidence and self assurance was unshakeable, often to a fault. Not self effacing and certainly not a yes-man to which my superiors would attest, I knew that to swim against the tide made me stronger. I had no interest in having any part of a popularity contest, believing complacency breeds mediocrity. I fought hard for what I believed, often entered when the sign said "keep out." While aware that perfection is for heaven I still hankered for it, was hard on myself when I failed.

While possessing a strong social conscience I am not a socializer and declined many invitations to join service clubs and fraternities. Required to be at work-related social functions, I disliked their hail-fellow-well-met, insincere back-slapping and frivolous small talk. These affairs bored me. An evening with friends was much more appealing.

Mother stimulated my curiosity to travel and explore. She believed being in a strange place was almost as good a way to get an education as in the classroom, so I have a strong wanderlust. From the time I was knee-high to a grasshopper she taught me the importance of goals, priorities and deadlines, to do the easy questions first to give more time for the hard ones. If away when I came home from school, she put out a list of tasks to be done before she returned. "Play" was always at the bottom of the list. Expected to do my share, work was part of growing up and being part of the family team. She inspired me to believe much more is possible in our lives than our imaginings can conjure.

* * *

IT IS LIKELY the older I get the better I used to be, so I hope this nostalgia has not transformed black and white memories into vivid technicolour, heightened the thrills of success, deadened the pains of defeat. I have struggled to defeat selective-memory-syndrome, to not distort facts for the story's sake or describe events the way I would have liked them to unfold. The syndrome may have crept in but to the best of my knowledge all of what follows is true. I apologize for any errors or for taking credit for accomplishments of others or for causing anyone harm.

Many of these anecdotes were notated in my personal and work diaries with posterity looking over my shoulder. Some incidents were detailed when this idea germinated many years ago. Others are from saved newspaper clippings and office memorabilia. Former staff subordinates have filled-in some blanks. Other episodes were stored on the back shelves of my mind. Often the deeper I dug the more things became exposed, just when I thought the excavation complete. Stories of the past are so easily protected with a coat of varnish, only the good parts are told without exposing the warts. Most of mine are related in chronological order, but there are many places where topical sequence made more sense. Also, I have tried to simplify the technical jargon of electric utility industry to spare the reader having to understand watts, kilowatts, sectionalizing, pulled sleeves, synchronization or oil circuit reclosers.

Despite my fierce employer loyalty throughout my career, a disappointment has been SaskPower's indifference to this effort. In their head office delving for information in both 1995 and 1997, I approached the then president and also the public affairs vice-president about this snippet of personal history and its connection to the company's roots. Under the illusion they would be at least curious I was kidding myself. I did not want funding or any special favours, just their interest. None was shown, not even to glance at my story outline. I left the building each time with the feeling that I was being a nuisance.

But that was then, and this is now, and much of what I saw and experienced is now gone. My disillusionment with SPC is tempered by the dramatic changes the electric utility industry has undergone since my retirement. I do not understand the complexities of deregulation; abandonment of franchise rights; privatization initiatives; sharing of transmission systems to move power in and out of the province where the stakes are high and the dances are delicate; or shortfalls in generation capability to the extent that some utilities are now unable to meet their province's needs, flaunting the very purpose for which they were initially established. Then there is the matter of security and response plans in the event of terrorist threats and criminal activities. Notwithstanding, history often resurfaces in strange ways. More often than not, changes inflicted by political invertebrates who, more likely than not, compromised their principles to grab power in the first place, are long-gone by the time taxpayers and customers reap the whirlwinds of their brief flirtations with power and their ill advised manipulations.

* * *

IN 1983 I MET Betty MacIntyre. She vowed after her divorce, as had I, to not remarry, but we did. Her unflinching support of my passion to keep my head busy, like writing these pages that peer into my working-life's window, has been a huge encouragement, as is her patience to repeatedly listen to these anecdotes. I am deeply grateful for the gamble she took with me and for the love and happiness we have found together.

1951 – 1952

In the Beginning

MY 35-YEAR CAREER in Canada's energy industry started 10 days after my 26th birthday in 1951 when I entered the personnel office of the Saskatchewan Power Corporation (SPC), the Wheat Province's electric power utility, and applied for work. Their head office was in the austere, sandstone, three-storied structure at 1739 Cornwall Street, a building I'd been in 10 years earlier during high school days with my pal, Pete Thornton. His father, Louis A. Thornton, was the first commissioner of the Saskatchewan Power Commission from its 1929 beginning until 1945, before the provincial government killed and buried it in 1949 and then birthed a more autonomous Crown Corporation. While we waited to see Mr. Thornton in his second floor office, usually for Pete to finagle keys to the family Buick or to negotiate an allowance advance, we cooled our heels in the overstuffed, swivel, leather chairs of the adjacent boardroom. Was this prophetic for my future?

Earlier in April, I had abruptly resigned from Steele Briggs Seeds, my employer for three years, after requesting a raise in pay from the manager, George Boyd. Pay increases were never given automatically or without asking. His crappy comeback to what I thought a legitimate request was 25 cents a week. Big deal! I was insulted and told him so when I quit. I deserved more and he well knew it. They had seemed pleased with my work as their representative in the eastern half of

Saskatchewan the first year I was employed by them. The next year I was promoted to be their British Columbia coastal salesman. During my two years in that territory I had increased business more than tenfold, an impressive amount. Boyd tried hard to dissuade me from quitting and avowed, "Dave, you have a promising future in the seed business and I believe you are making a career mistake to leave Steele Briggs. I will try to get you a little more money." His exhortations were unconvincing because their Toronto headquarters had such a tight-fisted attitude towards fair-pay, particularly for western hay-seeds, coupled with an absence of any employee fringe benefits, apart from a two week paid vacation, which was conceded to staff only because it was imposed by provincial legislation. Two of my confreres, journeymen seeds-men, trained in Britain, with many years of loyal service, were paid little more than me, a trainee. I earned $147 a month, including my raise, and worked a 44-hour, five and a half day week.

While unemployed for a couple of months that summer, I had contentedly busied myself doing odd jobs on my parents' acreage on the banks of Wascana Lake. This was on Regina's outskirts in what is now the Wascana Centre Authority. During that time I became engaged to Jean Reid, a nurse, and we planned to marry that September. She was a bit anxious over prospects of wedded life with a spouse not gainfully employed. Her mother was even more anxious.

My interest in working for SPC was piqued by a Regina Leader Post newspaper article about farm electrification. The Wheat Province lagged other Canadian jurisdictions in providing electricity to its rural citizens. Saskatchewan's governing political party, the Cooperative Commonwealth Federation (CCF) was on the verge of launching an accelerated program, wanting to provide strong leadership to help farmers get this much sought after service.

After grade 12 graduation in 1942, while waiting to age enough to volunteer for military service in the Second World War, I worked for a year as an electricians helper with Tibbitts Electric in Regina. Following my 1945 honourable discharge from the Navy I was enrolled for two years at the University of Saskatchewan's College of Agriculture, but did not graduate. I had enjoyed my work with seed-growing farmers at Steele Briggs and wanted to continue that association. So my work experiences and post secondary-studies reinforced the legitimacy of my interest in farm electrification for the agricultural community.

My personal employment references were two letters from Regina friends Al Gillespie, owner of Queen City Florists, and J.E. (Red) Adair, principal of Scott Collegiate, plus my gold wrist watch awarded to me by the Regina Leader Post newspaper in 1941 and engraved, "For Eighteen Months Perfect Service." I was proud of that achievement and it also seemed to impress SPC's R.J. (Bob) Waller, their prematurely grey-haired personnel manager. He seemed to have more than a casual interest in me and explained the Corporation's union contract with the International Brotherhood of Electrical Workers (IBEW.) Then he said, "There are no vacancies in the Farm Electrification Branch right now. Employees already on staff anywhere in the company have the first opportunity to fill a vacant position. If there are no qualified applicants for a particular vacancy we then hire externally." While disappointed with this news it was obvious there was no other alternative for me but to start at the bottom of the ladder and hope for an opening in the area of my interest. Then he continued, "But we do have an opening right now for a junior accounting clerk needing a grade 10 education." While I had no interest in accounting work of any kind I did not expose my hand.

I had done well on an IQ test and Waller suggested I complete a Kudar work preference profile. Having been alerted to the whereabouts of the only available job and knowing that to get into work that was of interest to me I had to get hired. So, without life jackets, I threw overboard my personal attributes of truthfulness and integrity and fibbed every question about accounting as being my work preference. This act of intentional deceit for personal benefit bothered me as I knew my personality was people-oriented. When the test was scored Waller looked pleased and said, "Accounting seems to be right up your alley." Without missing a heartbeat I said, "I'll take the job." He called the accounting manager, Ken Allan and reported, "Ken, I've got just the right man for your junior clerk position in Accounts Payable. You'd better come and meet him."

Four days later, on June 25, 1951, SPC's employee 523 reported for work in the Accounts Payable Section of the Accounting Division. My salary was $145 a month. Working conditions and benefits, all foreign to me at Steele Briggs, included a 37.5-hour, five-day work week plus fringe benefits of time and a half pay for any overtime work, a jointly funded employee-company retirement pension plan, automatic pay increases tied to work time and a satisfactory performance, paid sick leave that accumulated at the rate of a day and a quarter per month, two daily work breaks of 15 minutes each and a three-week paid vacation.

Ten years later, while climbing my success ladder when I was then SPC's Public Relations Superintendent, I skipped a team at the Tartan Curling Club. Bob Waller was my lead player. When socializing after games we often bantered about the day he hired me and, knowing of the only vacant position before I fudged the preference test, he had not gleaned where my work interest really lay.

* * *

I HAD NO FURTHER social or work contacts with Ken Allen after my stint in accounting. However, we became reacquainted in Victoria in 1986 after my retirement. Living alone and lonely after his wife's death, Ken visited us often as we lived close-by, and a friendship developed until his sudden death at 84, on March 23, 1999. We often spoke of our respective work experiences. He read early drafts of this memoir, fascinated, surprised and curious about my adventures in the early days of farm electrification, unaware of the details that went on behind the scenes of how the job was done, as his 42-year SPC career was with blinders on, behind an accounting desk day after day in

Head Office, paying bills and the payroll, accounting for inventories of materials and supplies, keeping the company books.

Accounts Payable

SPC's ACCOUNTING DIVISION and its Accounts Payable Section were crowded into the first mezzanine floor at the back of the Head Office building. The whole place was jam-packed with workers and office equipment as staff levels struggled to keep pace with the explosion of new customers and their growing demands on the province's electrical system.

SPC was quickly becoming Saskatchewan's biggest company and was a large purchaser of goods and services needed in every corner of the province, from the most complex electrical components to generate, transmit and distribute electricity to meat and potatoes for power line construction crew cook-houses. Basketfuls of purchase invoices, hundreds daily, landed on my desk, were manually numbered, recorded and matched with a Material Receiving Report that had originated with the employee who had ordered the goods or services. The two documents were the basis for payment to the supplier. The work was uncomplicated, repetitive and mundane, and did not challenge me by any yardstick, apart from the daily need to process large volumes of paper. There were however benefits to my humdrum task. I learned about the organization, its business terminology and technical terms of the trade, and became acquainted with staff in all areas of the company, many of whom were, or later became, senior managers and executives. I privately set myself a one year target to get out of accounting and into farm electrification work.

After a few weeks on the job I cranked up my courage and boldly went to see C.E. (Charlie) Smith, Superintendent of the Rural Electrification Branch, about my desire to work in his shop. With my background and declared interest he seemed impressed and concluded our pleasant conversation with, "Dave, I'll call you the minute a vacancy opens up."

* * *

SOON AFTER STARTING in accounting came my first encounter with a trade union, the IBEW, for whom my membership dues were a monthly pay-cheque deduction. It happened, of all places, in the men's washroom. As I bent over a sink to get a drink in my cupped hand from a cold water tap a husky voice from behind surprised me with, "Aren't there any drinking fountains in this place?" "No, none that I'm aware of," I innocently admitted. "Well! The power plants have them all over the place. I'll get that fixed," he growled authoritatively and, with a cigar clenched between his teeth, introduced himself as Max Clayton. He worked in one of the Corporation's power plants and was a IBEW executive member at a contract negotiation meeting across the hall in the boardroom, where I had mischievously twirled in the overstuffed chairs 10 years earlier. Soon after, plumbers arrived. The fountains appeared. I was impressed. Steele Briggs staff sure needed a union for protection from its management. Some years later Max and I became good friends.

* * *

AFTER THREE MONTHS my probationary period expired and I automatically got a five dollar raise. "Already yet," as my Ukrainian friends used to say, I was then making more than when at Steele Briggs. At the end of six months another increase of the same amount. My personal dream at the time was to climb the ladder at SPC and earn $250 a month, which I thought an almost unattainable whopping sum.

* * *

ON SEPTEMBER 22, Jean and I were married. In addition to this lifestyle adjustment, I volunteered for two work related activities, the Head Office Social Club and Club 1000. Both were useful contributions to my knowledge of business principles, supervisory skills and the role of basic management in a large organization. As well, both tasks helped keep my sanity in check as I waited for Charlie Smith's call.

Social Club

THOSE WITH WHOM I shared coffee breaks enticed this new-kid-on-the-block to become secretary treasurer of the Head Office Social Club. Their assistance with bookkeeping, if I had bitten-off more than I could chew, was assured. Sucked in like the morning milk, I was elected at the club's 1951 annual meeting.

Club membership dues for the head office staff was 50 cents a month, made by payroll deduction by the company and deposited in the club's bank account. SPC also provided basement space for a canteen, along with the tables and chairs, and paid for janitorial services and the utilities of heat, light and water. With a constitution, elected officers and various committees, the club operated the coffee shop in what had formerly been a courtroom for the Royal Canadian Mounted Police, the previous owner of the building when it was known as the Provincial Police Building. This space had vestiges of greater intrigue, like the magistrate's bench and prisoner dock and, before my time, even the jail holding cells and jury box were in place.

I sold bundled cardboard purchase tokens to the building's staff. One dollar bought ten. They were used instead of cash, for reasons of security, to purchase sandwiches, soup, cigarettes and tobacco, fruit, soft drinks, chocolate bars and other confections. Coffee was 10 cents a cup, including refills. I paid for the canteen supplies, salaries of the two staff, did the banking and kept the books. The canteen profits and membership dues funded head office staff social events – a children's Christmas party, a summer picnic, two or three dances a year, gifts for staff about to be married, or who were in hospital, or who were about to leave SPC. With the money, we also made donations to Regina charitable organizations, so staff could legitimately contend they "gave-at-the-office." My salary was five dollars a month.

My position required the supervision of the two ladies who worked for us in the canteen. The older, Maggie, the more senior of the two, was a feisty, single mom with a grumpy disposition. I often found myself in a referee's role when the two of them got at logger-heads, usually accompanied with Maggie's threat to quit. I did not trust her and suspected she stole canteen supplies, mainly cigarettes. Before her arrival one Friday morning I took inventory, repeated it that night after closing. There were shortages. Bill Reed, Club President, and I agreed she had to be let go. I mustered courage to do so over the weekend, then on Monday morning I happened to learn that she despicably pimped for her teenage daughter. My plan to fire her Monday afternoon at quitting time was fortuitously precipitated by another argument after lunch with her co-worker. When the predictable, "I quit," threat was made I had her sign a letter to that effect, recovered her keys and ushered her out the door amid language befitting a hooligan. I had ably survived my first managerial decision and its execution.

The major social club sponsored affair was an annual formal Christmas dinner and dance at the Hotel Saskatchewan. It was a prestigious event in 1951, attended by several hundred employees, spouses and friends crammed into the hotel's Ranch Room and an adjoining ballroom. A photograph of the occasion even appeared in SPC's Annual Report that year. Head table guests and spouses congregated in an upstairs hotel room where winter clothes were left, corsages were pinned on the ladies and cocktails were served as they socialized with the club's executive. As an Accounting Junior Clerk I was enchanted to be hobnobbing with the province's most powerful political pundits.

It was my first face-to-face encounter with Premier T.C. (Tommy) Douglas, SPC Board of Directors Chairman J.A. (Jim)

Darling, several board members including Woodrow S. Lloyd, Minister of Education, Allan Blakeney, the board's secretary and at the time not in any elected office, Jack Tomlinson, SPC's General Manager, and the other senior executives and their wives. Interspersed at the banquet head table, which ran the full length of the Ranch Room, sat the club's executive and their spouses, the premier and two future premiers, the board members and SPC's management. The very presence of that number of the province's political leaders signalled the importance they felt about the Corporation's role in Saskatchewan's future. The remarks of the various speakers pervaded the vision about building a province-wide electrical system that would serve all citizens, particularly the thousands of unserved farms. Douglas generously laced his remarks with humour in his inimitable style. One anecdote was about his wife's comment as they drove down a highway one day, when she apparently asked, "Why do the power men run up the poles as we approach, was it because they knew you are such a bad driver or that it is the premier behind the wheel?"

I was particularly inspired by Douglas's passion and challenge he threw out that night to somehow get electricity to every Saskatchewan farm. His fervour whetted my desire to get into the Farm Electrification Branch.

My social club's Secretary Treasurer term ended in the fall of 1952. Unlike Tommy Douglas, I did not seek re-election.

* * *

I INFORMALLY MET Tommy Douglas many times during his premiership. While he did not remember me by name each time, he never needed reminding that I worked for SPC. I admired and respected him, believed him to be an astute politician and a leader of integrity who stood fervently and passionately for what he believed. It is rare breed to find both qualities in one person. He and my Father were on a first name basis.

Club 1000 and the Power Credit Union

IN APRIL, 1950, before I started with SPC, a handful of enterprising staff organized a financial club for saving money from which other employees could make loans, security being the borrower's character. These small loans, all of the payday variety, had to be repaid in full at month-end. This was conceived by six friends one morning at coffee in the canteen when Leo Kobayashi, a draughtsman, needed five dollars to tide him over a weekend. None of the other five could make the loan. However, each could put up a dollar. One of them, Henry Kilarski, was appointed treasurer on the spot. Leo got his loan, to be repaid on payday plus 25 cents interest, a usurer rate if there ever was one. Their idea mushroomed into elected officers and formation of "Club 1000." The objective was to have $1,000 in member deposits by year-end, hence its name. A year later $719 was held by 29 employees, who received an annual dividend based on the size of their invested savings, profits earned from interest on the loans that had been made and bank interest from the funds held on deposit. As with the Social Club, Club 1000 was managed by volunteer head office staff, contributing their time outside normal working hours. The Corporation contributed meeting space for the board and committees throughout the life of PCU, wanting to add to the benefits its staff would receive from this self-help movement.

In 1952 the manager of the Bank of Montreal, holder of the club's current account, queried the volume of transactions. When given the details he told Kilarski a deposit and loan operation outside of the Government of Canada's charter for financial institutions was unlawful and contrary to the Canadian Bank Act. He suggested a credit union would be more appropriate. On August 21, 1952, the Power Credit Union Limited (PCU) was formally created. By year-end 44 charter members, including me, had deposited a total of $5,006 in assets. A five per cent dividend on shares was paid that year. Interest on loans was one per cent per month on the unpaid balance and a year-end patronage refund was made to all members.

* * *

CREDIT UNIONS WERE not new to Saskatchewan. The first community types were chartered as early as 1937, when the Credit Union Act was passed by the province's Legislature. But an industrial one expressly for employees of a company was uncommon, PCU among the first.

Credit unions were based on the Christian ethic of the Golden Rule, to help one another and to be your brother's keeper. Criticism of money lenders, usurious interest rates and fraudulent use of funds entrusted for safekeeping have been recorded long before the birth of Christ and continue to this day. During the Great Depression the calloused actions of chartered banks, mortgage, trust and insurance companies, with their heartless foreclosures against farmers, spawned interest in the movement. A tale told in those dark times was about a bank loaning an umbrella in a dust storm but demanding it back when it started to rain.

The credit union principle was, "Not for profit, not for charity, but for service." Managed by its own people it functioned on the basis of one member, one vote, regardless of the number of shares that person had on deposit. As compared to a bank where a customer is only a purchaser of a service controlled by investors, the users of credit control a credit union. Roy Bergengren, an American writer and philosopher said, "That money should be used for the benefit of those to whom it belongs." Those who pooled their meagre resources for mutual benefit found a special kind of magic in a credit union. SPC employees were not exceptions. I found this magic contagious because of the earlier hardships in the Dirty Thirties my parents and I had endured, caused, in part, by intransigent financial institutions and indifferent governments. Hard times make cooperators of us all.

SPC was supportive of PCU and from the outset accommodated employee payroll deductions without cost. This was an encouragement for staff to save as they didn't spend what they didn't have. There were benefits to SPC as, over time, PCU salvaged many employees who became indebted well over their heads. It was then Corporation practice to dismiss an employee when a wage garnishee was served from an impatient creditor. With management support PCU counselled the tardy paying or overspending employee, encouraged debt consolidation with a single loan which in turn paid off the creditors. SPC averted the loss of staff in whom it had a considerable investment to train and upgrade. In later years when PCU was able to handle mortgage loans, Corporation field staff moving from one small community to another had their financial needs met easier than from other conventional and more reticent mortgage and lending institutions.

In 1952 I was elected the first chairman of PCU's three member Credit Committee, along with Ruby Drummond and Bob Byzick. We scrutinized every loan application. The Credit Union Act required, "All loans granted shall be for provident and productive purposes." At the 1953 annual meeting I reported our approval of 111 loans for over $21,000. A few had been declined and we had incurred our first bad loan, ironically by Bert Denys, the employee I had by then replaced in the Farm Electrification Branch. More will be said about this shortly. That loan was collected from the cosigner who was also an SPC employee. My year-end report urged members to be cautious for whom they underwrote a loan.

PCU was associated with the Credit Union National Association (CUNA), in Madison, Wisconsin. They provided life insurance coverage on all loans, shares and deposit accounts. This was particularly valuable to our members as many were construction workers building power stations, electrical substations, high voltage power lines and natural gas high pressure pipeline facilities. Accidents inevitably took the lives of one or more employees every year in spite of the Corporation's aggressive Employee Accident Prevention Program. If the victim had a PCU loan or mortgage at the time of the tragedy, CUNA paid off the debt and deposited an equal amount in the deceased's share account. It gave me great personal satisfaction to be part of a movement which looked after the financial interests of its members, even unto death. When our manager reported on his visit to a grieving family and had told them their loan or mortgage had been paid in full and an equal amount deposited in their account, I felt good.

* * *

OVER 20 YEARS LATER, in 1974, and again in 1975, when I was Consumer Services Division Manager, I was elected PCU's President and a member of the Board of Director's Executive Committee. My office was the venue for meetings of the board, its members from both SPC's unionized staff and management ranks, who headed the various committees of Supervisory, Credit, Promotion, Member Services and Personnel. We were a close knit group with a common cause.

At one board meeting we considered an insurance plan for staff and board members. A Saskatchewan bank manager's family had been kidnapped by thugs for ransom, which was delivered, but with police cooperation the gangsters had been apprehended. The Credit Society recommended credit unions protect themselves from similar threats. I felt it a sad commentary on society's changing times when we agreed to pay the insurance coverage premium.

By 1975 PCU membership exceeded 4,000: employees, superannuates and their families with assets of over $10 million. That year more than $3 million had been loaned to members. It was a big-time operation.

In 1976, Vice President Don Gilmour, Armand Pinel,

PCU's manager, and I negotiated our amalgamation with the Telephone Employees Credit Union. It had 458 members and $500,000 in assets. This merger enabled us to open a full-time downtown office in Saskatoon, a benefit to our members there, in office space contributed by the Saskatchewan Telephone Corporation. Full-time staff were then able to replace Barbara Hodson, an SPC secretary, who up to then was a part-time PCU employee performing the duties from her office desk. A few PCU members objected to this accord, felt we would get too big and lose sight of the original goals, and that credit union principles were best applied in small groups. The merger made sense to me as our operational costs could be spread over a larger membership base. A secret vote by members of the two organizations agreed with the amalgamation by a wide margin.

PCU had a remarkable growth. By 1976 with a staff of 15 the membership enjoyed a full range of financial services, each introduced when conditions permitted, often not without the pains of growth – full chequing service, financial counselling, computerized account transactions, lines of credit, mortgage, equity and consumer loans, safety deposit boxes, payroll deductions for purchase of Canada and Saskatchewan Savings bonds. Members were encouraged into special savings, estate, retirement and endowment plans, the preparation of wills and estate planning with help from the Cooperative Trust Company.

PCU's 25th anniversary celebration was planned for August 28, 1976. As part of those plans I recommended, and the board agreed, to have PCU's history researched and written. Grace Lane, a Regina writer and an acquaintance of mine had earlier written *Brief Halt at Mile 50* to commemorate the 50th anniversary in 1975 of the United Church Of Canada. We contracted with her for $5,000 to write PCU's history. She conducted interviews, undertook research, obtained photographs and produced a 45-page book, aptly titled, *Minding Your Own Business*. Completed in time for the anniversary party, a copy was given to each member at that dinner.

* * *

PCU is now gone, swallowed-up in a further amalgamation with the Provincial Association of Government Employees Credit Union, which still has a branch in the nearby Financial Building which services nearby SPC Head Office staff.

I still have feelings of pride in what I was able to contribute to this unique, self-help financial movement.

1952 – 1954

The Father of Farm Electrification

JUST AS MY self-imposed one year deadline was about to expire to escape from accounting, Charlie Smith called me, as he had promised months earlier, and said, "Dave, one of our farm electrification clerks, Bert Denys, has just resigned and a replacement for this position will be advertised in a few days. Be sure to watch the bulletin boards and put in your bid as, with the seniority you now have, I expect you should have a good chance to get the job. I hope you do, as I would like to have you on my staff." I applied and when the bids closed I had the longest whiskers and was appointed on August 11, 1952. In addition to leaving a job in which I had no interest to one that did, it paid more. I had just climbed up one small step of my ladder. Steele Briggs had long-since gone out-of-mind.

* * *

CHARLIE SMITH'S CONTRIBUTION to the electrification of rural Saskatchewan has not, in my opinion, been publicly recognized so, before continuing with my story, which would not have happened without his influence, I want to pay tribute to him with a small remembrance, which he so richly deserves.

In 1952, Eleanor Stanton, the first full-time editor of SPC's employee magazine, *HiLines*, profiled Charlie in an article. It impressed me even before I knew him. She wrote, "Works with his employees, not over them." Her observation characterized his demeanor with his staff. He influenced my working career more than any other person. For 10 of my first 15 years with SPC, he was my supervisor.

Called Charlie by those who knew him, occasionally Chas, the only exception I knew of was David Cass-Beggs, SPC's General Manager from 1955 to 1964, who, in his English accent, called him Charles with emphasis on the "a." Charlie was a trusted confidant of Cass-Beggs throughout his 10 years at SPC's helm.

As with many societal changes which make this world a better place, there are those, not the conquerors described in history books or the icons of industry, but rather the teachers and role models, often out of the limelight, who provide the foundation, the backbone and the texture of society which help to enrich our lives. Charlie was such a person, a builder who was ideologically driven.

While history has recorded Saskatchewan's rural electrification miracle and others may claim different people were responsible for this accomplishment, the one who trudged in the muck of the trenches and sweated out the day-to-day issues was Charlie Smith. He was, I believe, the unquestioned father of rural electrification. He was its heart and soul. He developed the plans, policies, regulations and procedures, organized the farmers so mass production techniques could be applied to build the lines, managed the capital funds so the most could be done with the available money, and garnered overwhelming support for the task from tens of thousands of farmers.

Chapter 11, entitled "Operation Stubble Jumper" in C.O. White's book, *Power for a Province*, a history of Saskatchewan Power, quotes several of Charlie's memos about farm electrification. He often wrote directly to Jack Tomlinson, General Manager, even to Jim Darling, Chairman of the Board of Directors, almost bypassing his superiors. While perhaps not good protocol, they trusted him. I now know how suggested solutions to policy issues Charlie's staff recommended were so quickly approved.

When Charlie spoke of politics, his passion for the human values of the CCF party was unmistakable. He was an avowed, loyal, card-carrying CCFer. In 1944 while in Tisdale, he helped re-elect the local candidate, J.H. (Brock) Brockelbank. In Doris French Shackelton's 1974 biography, *Tommy Douglas*, she writes about that political campaign when the incumbent premier, seeking re-election, spoke of the need for electricity on Saskatchewan's farms. Knowing that the farmers wife would benefit the most from this service, he appealed to voters to imagine, "Farm women working without the benefit of electric washing machines or stoves." Charlie once told me about that June 15, 1944 election night. Several like-minded employees huddled around a lunch-room table in the diesel power plant they operated. The radio blared the election returns over the rumble of the engines, pumps and generators. Not only was Brock re-elected, but the party would, for the first time, with 47 seats, form the new provincial government with only a five seat Liberal opposition. Ten years later the event still brought a glint to his eyes as he recalled their shouts, "We Won! We Won!"

The expression, "The first to arrive, the last to leave and took the shortest lunches and coffee breaks," best described his work regimen. It captivated his full attention. He drove himself hard to his own and his family's detriment. This enthusiasm and work ethic infected those around him. It demanded the best. I too was caught up in his intense passion and zeal, had the sense I was part of something special and really worthwhile.

Charlie was kind, considerate and compassionate; approachable and supportive; humorous and humble; empathetic and fair; as much at ease with the premier as he was with a modest quarter section farmer or a child of one of his staff. The writer, Elbert Hubbard best describes my respect for Charlie, "If you work for a man, in heaven's name work for him! If he pays you wages that supply you bread and butter, work for him, speak well of him, think well of him, and stand by him, and stand by the institution he represents."

Charlie's visible trademark at the office was the green ink in his fountain pen that commanded, at least for me, special attention when that coloured signature appeared on a letter or memo.

Charlie, born in Iowa, came to Canada as a child with his parents, who settled on the banks of the Saskatchewan River, 30 miles north of Swift Current. Their home was the historic "Goodwin House" at Saskatchewan Landing. Mr. Smith ran the ferry which plied the river now replaced by the Saskatchewan Landing Bridge and ran a small store and market garden. At 20, in 1930, with his high school education under his belt, Charlie joined the Saskatchewan Power Commission and worked in their Swift Current power plant. In 1939 he went to Maple Creek, in charge of that diesel power station and maintenance of the power lines that served nearby customers. Later that year he became Watrous Assistant District Superintendent and, in 1942, District Superintendent at Tisdale. Despite difficult material shortages caused by the Second World War, he expanded these small stations and the local lines which emanated from them to serve a few other small communities and several nearby farmers. In 1949, Roy Sarsfield, Sales Manager, brought him to Regina as Farm Electrification Superintendent.

The Liberal party was elected in 1964, ending a 20 year CCF reign. By then I had followed in Charlie's footsteps and was SPC's Public Relations Superintendent. The Liberal's were well aware of his long-time political allegiance but respecting his pioneer rural electrification work they did not, to their credit, make a public example of him to suit their political agenda. He maintained his title and dignity, but some of his work duties diminished over time, reassigned to others and his influence in management decisions lessened. I felt sorry about this, often lent a sympathetic ear. He was then a group head (vice president) and continued to attend meetings of the new Board of Directors, and quietly expressed to me his concerns about some of their decisions that were leading the Corporation along different paths. But he never betrayed them or gave any reason to suspect his fierce SPC loyalty.

In 1969 when I was Consumer Services Division Manager, Charlie decided to retire after going on a province-wide tour to visit his many associates, both in and outside SPC. On the evening of September 17, just as he had started that tour, at age 59, he died of a coronary thrombosis. He had collapsed the day before in Nipawin while visiting an old friend, District Operator Len Hewer. I believe his untimely death was caused from stress associated with hard work and his heavy smoking. Charlie's second wife, Jean, his daughter Brenda, and his best friend, George Busse, were with him when he died. His second daughter, Deanna, arrived too late in spite of my best efforts to get her there earlier on a chartered flight from Saskatoon, where she had flown from her Winnipeg home.

Charlie's family asked if I could arrange his funeral service in Regina's Lakeview United Church where he had walked down the aisle and given both Brenda and Deanna away at their weddings. In our many private discussions Charlie made no bones about his being an atheist so it was with some trepidation that I approached Lakeview's minister, Dr. Allan Martin. He understood and was co-operative, even to my request to allow Father Matthew Michael, a retired Roman Catholic priest from St. Benedict, to deliver the eulogy during the protestant service which Dr. Martin would conduct.

Father Matthew and Charlie were respected friends. Matthew was an early and committed advocate of farm electrification and regularly preached of its merits in his homilies to his St. Gregor and Meunster parishioners. A determined cleric, he had pushed the interests of his flock towards a better lifestyle, about 150 of whom were one of the first large blocks of farms to receive electricity in 1949 as a Rural Power Cooperative. When his diocese's Bishop transferred him to another parish, those he had assisted to get electricity gave him a new Chevrolet, a significant gift for the times. The district to where he was moved was not electrified anywhere near as well as the one he left. Charlie jested he was going to ring up the Pope and request the Vatican keep Matthew in one place and out of our hair as we knew his unrelenting pressure to help his new congregation was just around the corner.

Dr. Martin, whose protestant theological background was

quite different than Father Matthew's, did not want any discourse from the pulpit that might be controversial. To ensure his role was to eulogize the value of Charlie's work and his character I lunched with Matthew the day of the funeral to talk about the afternoon service format. In his homage, the priest praised Charlie's attributes as a man, the significant benefits he had given to Saskatchewan's rural people and the heavy burdens electricity had lifted from the shoulders of farm families. His remarks, generously flavoured with humour, which Charlie would have enjoyed, brought forth several bursts of laughter, an emotion not usually experienced in funeral services at that time. It was a sad day for me. I had lost a respected friend, as did many other staff and tens of thousands of Saskatchewan farmers.

*   *   *

FARM ELECTRIFICATION WAS foremost amongst all the economic and social programs carried out in the two decades of CCF government from 1944 to 1964, an achievement that can be attributed to the outstanding leadership of Charlie Smith. He, more than any other, was the mahatma of that miracle! I know, because I was there. I saw it happen.

The Rural Electrification Act

SASKATCHEWAN WAS IN the backwater compared to other Canadian provinces, in so far as addressing the issue of the electrification of its rural area. By 1949 the others were in varying stages of completion. This last place finish was the result of a widely scattered farm population, the huge investments required to build the intricate web of power lines and related facilities needed to connect them, the lack of political will up to that time to address the question, and the unwillingness of the province's privately owned power companies to even think about such unprofitable investments. The investor-owned utilities served only the larger communities, enabling them to earn maximum returns on their investments. Hooking up smaller towns and villages, let alone individual, isolated farms, was not a profitable business, unless the customers paid all the costs to get the lines built. As far as the companies were concerned farmers could stay in the dark.

Manitoba and Alberta were well along with their electrification programs. The Keystone province's publicly-owned utility subsidized all the power line cost for their farmers. But this financial burden was easier for them with only two thirds of the number of Saskatchewan farms compressed into one third of the geographic area, making their investments per farm significantly lower. The condition of distances and costs in Alberta were similar to Saskatchewan, but its investor-owned utilities met the problem by simply requiring rural customers to pay all the line costs to get service, in some cases well over $1,400 each, so there was little risk involved for the utilities.

Saskatchewan did not have an enunciated policy, let alone any idea how the job could be accomplished. In its 1928 report, the Power Resources Commission all but ruled-out farm electrification as a viable scheme. This was followed a few years later with the silly suggestion the government assist farmers to buy individual small lighting plants rather than build power lines to their farms.

* * *

TO HELP UNDERSTAND the magnitude of electrifying rural Saskatchewan, the reader, at least those from afar, should have some appreciation of the province's size. It is 761 miles from its southern border on the 49th parallel to its northern boundary at the 60th parallel. Her widest point, east to west, along the USA border is 361 miles and 277 miles at the northern boundary. Its northern half is in the hard rock Canadian Shield and is, relatively speaking, almost uninhabited, primarily boreal forest dotted with hundreds of lakes with little arable ground cover. The southern half that is under cultivation, from which its nickname the Wheat Province is derived, covers over 125,000 square miles. By 1949 it was believed there were about 120,000 individual farms so, on average, each farm's size was then about one square mile.

As the average acreage per farm in Saskatchewan is the greatest of any other Canadian province, it gives her the unenviable distinction as one of the most isolated places in North America where so few people are scattered over such a widespread area. Wallace Stegner describes this sparse population and its loneliness for early settlers, coupled with its harsh weather, in the opening paragraphs of his book, *Wolf Willow*, "This is no safety valve for the population explosion, no prize in a latter-day land rush."

In 1872, to foster settlement of the prairies, the Ottawa government enacted the Dominion Lands Act which dictated the lineaments of the West. Canada's land survey system copied that of the American's. Survey parties spread out across the land

25

barely ahead of the the settlers to create an orderly pattern of townships, each one being six miles square, or 36 square miles. Each square mile was called a section, 640 acres divided into four quarters of 160 acres, each one half a mile square, called a quarter section. These were the parcels that were allotted for a 10 dollar fee to the homesteaders that came from eastern Canada, the United States and Europe. The survey also provided for road allowances, 66 feet wide, a mile apart for those running north and south, and two miles apart for those going east and west.

Many early settlers did not endure their wild-west venture. Some departed after experiencing only one harsh winter, and abandoned or sold their land to neighbours who survived and expanded their acreages. Even in those early days there was not a settler on every quarter section so farms were isolated, one from the other, a lonely existence particularly for the farmers wife. Over the years as the amount of acreage on the average farm increased the distances between farm sites also expanded so that even todays sod-busters have to endure a survey system that inhibited the advantages of human community.

It would have been advantageous if the survey had copied the French Canadian system of strip farms with long land allotments running away from and at right angles to the rivers that were their highways. Applying this scheme to the prairies, with roads instead of rivers, farm sites would have been relatively close together. Such a system would have, over the years, saved fortunes, many times over, in the costs to provide and maintain municipal services. The big advantage would have been in the social benefits of settlers being able to live in closer proximity. Unforeseen at the time was the invention of telephones and electricity. Imagine the savings in line construction costs with a more concentrated population on the prairies. Such was not to be the case.

In any event, the governments land survey system and settlement policies complimented the Canadian Pacific Railway's (CPR) marketing slogan as "The World's Greatest Travel System" which they promoted throughout Europe over 100 years ago that promised immigrants coming to Canada "a land of milk and honey." Their passenger steamships brought the eager settlers across the North Atlantic to Canada's east coast, there to embark on their trains for the trip further west. The homesteaders required consumer goods and machinery to be transported from the eastern factories and, of course, cattle and produce needing to be shipped east to market. So the CPR's trains profited, both coming and going, even to the extent their gouging freight rates for goods transported west were about half the cost to ship them east, an unprincipled business strategy that for years has contributed to the unrest and ongoing dissatisfaction of the Agrarian West with the Industrialized East.

In addition, the mercenary interests of the CPR were satisfied even further with overly generous free title for millions of acres of land as compensation for pushing the steel westward to deliver the goods from the east. Much of the land given the railway was the most fertile and productive. Less than a hundred years later the railway lines were abandoned, the steel tracks torn up and the CPR withdrew, having completed its corporate plunder of the prairies.

My use of the word "homesteader," which customarily describes prairie settlers, implies they were the first occupiers of these lands. Certainly they were new to the prairies and to the quarter sections legally granted them under authority of the Dominion Lands Act. As they settled into their new surroundings believing they were its first occupants, they erected homes, barns, schools and churches, cleared the trees, burned the stumps, broke the land then sowed and reaped the crops, expecting to profit from their hard labour. But they were not the first occupants, by far. This distinction belongs to aboriginal people. Indians and Metis, prairie nomads for thousands of years with their forms of leadership, government, law and culture, were there long before the white mans' face emerged. They roamed and hunted on the plains and in the forests, paddled the rivers, ably survived the harshness of prairie weather, developed their language and customs always with greatest respect and love for mother nature. While that is another story, remember the nineteenth and twentieth century homesteaders were interlopers on this land now known as Saskatchewan.

This then, sets the stage for rural electrification.

* * *

IN THE SUMMER OF 1945, a year after Saskatchewan people elected their first CCF social-democratic government the politicians needed expert advice on electrical matters. They engaged a consultant, David Cass-Beggs, a University of Toronto electrical engineering professor, born and educated in England. Cass-Beggs gained their confidence as he was one of them and had been a member of the British Labour party before coming to Canada and joining the CCF, under whose banner he had been an unsuccessful Toronto riding candidate in an earlier Ontario provincial election.

Cass-Beggs's studies of the rather bleak Saskatchewan electrical horizon were the source of several innovative reports he authored which became master plans for the province's integrated power system. Two of them dealt with rural electrification, the second and more detailed document, dated October 1947, was entitled, *Report on the Development of the Facilities of the Saskatchewan Power Commission to Supply Electric Power to Rural and Farm Areas*. It paved the way for drafting of the Rural Electrification Act (REA) for a government anxious to bite the bullet and get on with the job.

The REA was proclaimed by the Legislature on February 1, 1949, and, at the same time, it created a new crown corporation, SPC, which replaced the Saskatchewan Power Commission. The earlier Commission had functioned much like a government department under close, some would say too close, control of the politicians. The new organization was intended to have greater autonomy over its operations than the one it replaced.

SPC inherited the Commission's 5,500 miles of power line, a conglomerate of 35 thermal and diesel power stations and a few thousand customers, including some 1,500 farms who had electrical service only because their farm yards were close to towns with a power plant or to a passing power line. Some larger communities not served by SPC or other private utilities had limited lighting service from power plants whose generators were turned by relatively small diesel or gasoline driven engines. These lighting plants were usually sideline operations of storekeepers, garage owners, machine shop mechanics or flour mill owners. Some served only other businesses in the downtown core and a few homes in close proximity to their plants. Customer costs for electricity from these small units averaged from about 25 to 30 cents per kilowatt hour (kWh) or more, as compared with 8 to 10 cents from one of SPC's larger diesel stations and even less from one of the more efficient power plants that produced steam from coal or oil to turn its generators. As well, the little operators only supplied 110 volts for lights and small appliances. Little wonder these small plant owners wanted to be relieved of their responsibilities and the people living in these urban communities wanted central station electricity without restrictions on their time of use or inability to use major appliances like electric ranges and water heaters.

Coincidental with the twin births of the REA and SPC was Charlie Smith's Regina appointment to start building a meaningful farm electrification program from those foundations.

The REA, characterized as a hesitant first step and a relatively conservative bit of legislation, set out the policy of the government's intention. It provided for farms to be connected in three ways: individually, or through a power district of seven or more farms, or by the formation of a rural power co-operative intended to handle groups of 100 or more neighbourhood farms. Each option had its own regulations and restrictions which, in reality, meant most new connections were to individuals or the small districts as few power co-operatives were able to organize. This, along with shortages of some power line construction materials in 1949 and 1950, reduced the number of farms actually served than what had been envisioned.

Then, the relatively small number electrified in 1950 and 1951 made it clear the REA needed revision to accomplish what Cass-Beggs had contemplated and what the government desired. By the end of 1951 only about 8,000 farms had electrical service, less than seven per cent of the province's total.

Universal electrification for the entire province needed a more assertive and efficient approach, rather than trying to accomplish it on an individual, hit-or-miss helter-skelter basis. The original REA policies meant only those close to existing power lines could afford the service. There were instances of farmers organizing power districts and rural power co-operatives that selfishly and intentionally left out those neighbours who happened to be isolated from those close together, as in doing so reduced the costs for the main group. Those left out would either pay exorbitant costs or do without the service. These methods were inequitable and troublesome to say the least.

There was some urgency to proceed with the daunting task of rural electrification for two reasons. First, the farm economy had significantly improved, buoyed by emotional waves of optimism throughout Canada that emerged after the Second World War. Farmers had an air of confidence and, along with it, more disposable income. Just as the tractor and combine had redefined farm work, this improved prosperity spawned the desire for electricity which would raise their living standards, increase farm productivity and reduce some of the hard manual labour, a condition synonymous with farming. Second, this interest and demand for electricity came at an opportune time. The CCF government was chafing at the bit to implement an array of social and economic programs which would not only stimulate the economy but provide the province's rural population with a better way of life. SPC met this challenge with a plan called "Area Coverage."

Area Coverage

THE CONCEPT OF Area Coverage (AC) was to simultaneously electrify many farms in a broad area on a project basis using mass production techniques to organize and build the facilities with a manageable system of shared costs. Power line construction work would benefit from the economies of scale with lower costs to both the farmers and SPC and as well the job would be done faster. For those farmers remote from power lines and with little expectation to ever get electricity, this plan gave them the same opportunity and equality of treatment as those who happened to live close to existing power lines. To develop this concept to the implementation phase the Corporation needed to know where each farmers buildings were located.

There was no source, in or out of government, who knew the location of farm sites or their exact number, speculated to be about 120,000 at the time. It was critical their locations be known so the power lines would, when built and in operation, provide adequate and reliable service in the most economical way. But how could the location of each set of farm buildings be reliably identified?

The first brain-storming notion was to drive all the province's rural roads and plot farmyard locations on maps, along with any impediments to power line construction. These barriers were lakes, rivers, ravines, valleys and large sloughs, but also the preponderance of bush – trees, shrubs, scrub and wild undergrowth. In the north this growth took the form of black and white poplar stands, often called aspen, some three feet or more in diameter and forests of conifers, lodge pole pine and spruce. Some farmers called it brush, others referred to it as bluffs, an Eastern Canadian description of a rock outcropping, but whatever its name, it all had to be identified for cost estimating purposes, then removed for power line clearances.

Murray Leach launched the field checking idea in an automobile armed with a map of the Rural Municipality of Argyle, RM #1, in the province's southeast corner, bordered on the east and south by Manitoba and North Dakota, respectively. This municipality is 18 miles square and contained nine townships, 324 sections of land. Even though the land survey plans of the Dominion Lands Act of 1872 had provided for road allowances this was no guarantee that roads existed.

Right from the outset Leach found some road allowances were not passable, as they had never been developed. Others were but primitive passages, nothing more than two hollowed-out ruts with grass and weeds between them, often hiding large rocks, hazardous for a car's undercarriage and engine oil pan. Some were only graded-up dirt and had yet to see any gravel. Rain prevented Leach's travel on muddy roads for one day and he returned discouraged after the first week. A revisit would be needed to complete just this first municipality. As this mapping by car could only be done in fair weather and with a few hundred Rural Municipalities (RM) and Local Improvement Districts (LID) in the province, this scheme was quickly abandoned. Another, better method was needed.

More staff brain-storming led to talks with the Saskatchewan Association of Rural Municipalities. They agreed it was reasonable for SPC to ask each RM to identify farmyard locations within their municipality. It was believed each of the RM's six councillors, most of whom lived in the divisions they represented, knew where their neighbours lived and with not much effort should be able to precisely mark them on the correct quarter section on a map, along with the location of schools, churches, halls, inland stores and post offices, power line construction barriers and, finally, road allowances which were not in use or that had never been developed.

Blank maps with instructions about the information needed and why it was required went to every RM and LID secretary, with the request that each of their six councillors complete their respective division, a township and a half, a total of 54 sections. The maps, on a scale of one inch to the mile, with each section divided into the conventional four quarters, showed existing power lines, the incorporated limits of towns, villages and hamlets, railway lines, highways, land survey correction lines, provincial parks and major geographic highlights like valleys, rivers and lakes.

Completed maps started to arrive from RM secretaries not long after they went out, from those anxious and serious about getting electricity. The data from most councillors was reasonably accurate when later checked on-site, but there were troublesome exceptions. Some thought every quarter section warranted a set of farm buildings. Others we later found were prepared in the comfort of the councillor's kitchen or from the corner of a beer parlour table where memories were not quite as accurate as they surmised. A quick glance at these rendered them useless. Replacements were sent out or a visit by one of the farm electrification field representatives corrected these deficiencies.

News of SPC's seriousness about a massive farm electrification program spread like a racy rumour at a Ladies Aid meeting. There was no need for any advertising to alert the rural areas that electrification was a possibility. Applications for service reached flood proportions. If the municipal mapping had not been completed, petitioners were told their requests could not be processed until their council returned the needed information. While approaching extortion, this diverted some of the immediate demands as it pressured the local government leaders to finish the mapping. Even then it was a long time before some maps arrived. A few never did.

The data supplied by the local elected municipal people was then transferred to permanent maps. The engineers laid out tentative line routes for the 25,000-volt (25 kV), three-wire distribution lines. These generally followed the provincial highway system where regular maintenance and trouble-shooting after storms would be easier to handle by the operations staff than if they were located off the beaten track. These lines served the urban communities along their route and were the backbone of the rural system as it was these three-wire lines that supplied the single-wire 13,800-volt (13.8 kV) farm lines. Every farm location supplied to SPC was shown as being connected to a future power line, the ultimate goal. Many were 50 or a 100 or more miles from an existing source of power. Their remoteness led many to think they would never see poles come over the horizon and across the fields to their places.

* * *

A SIGNIFICANT CONTRIBUTION that improved the economics of a rural electrical network was the innovative development in 1949 by W.B. (Baldy) Clipsham, SPC's Chief Engineer, of a 13.8 kV single-wire rural power line that used the earth in place of a return neutral or ground wire. Previously, rural lines were of 6,900 volts (6.9 kV) carried by two conductors. The new design of only one wire could handle not only more load and, therefore more customers, it could also be extended a greater distance from its source. The wire size of this new single conductor could be reduced because of its higher voltage. It was made with steel and aluminum instead of the more costly, heavier and larger diameter combination of steel and copper in use up until that time. This one wire mounted on pole tops removed the need for a second conductor attached below and on the sides of the poles. This meant a reduced cost for wire, insulators and poles as their lengths could be shorter and only 14 were needed for a mile of line. This new design was much cheaper to build, a mile costing only $650, some $200 less than the earlier two-wire system.

Clipsham's modesty kept him from applying for a patent for his invention, satisfied this contribution was just part of his job. He did receive some recognition though when his technical paper about the achievement was published in *Spectrum*, the flagship journal of the Institute of Electrical and Electronic

Engineers. This type of single-wire line is now in use world-wide.

* * *

EXPERIENCE HAD SHOWN that every farm in any district would not take electrical service when it was made available to them, even on a project basis. More will be told about this later. For purposes of system design and costing it was assumed that in the long-term 80 per cent of all the farms identified on the maps would ultimately be connected.

The needed capital funds were borrowed by the provincial government on the money markets in New York, Montreal and Toronto for each year's program. The funds were advanced to SPC to build the new lines and other facilities. Over the life of these assets the government guaranteed the debt but the Corporation carried it as a liability on its books and repaid the investors as the bond repayments fell due, plus interest. As these annual borrowings were limited, the expansion costs needed to be controlled so the greatest number of farms could be served with the available money. It was planned farmers would bear about 60 per cent of the single-wire rural power line cost with SPC carrying the balance plus the cost of the higher voltage three-wire 25 kV distribution lines.

The municipal mapping data gave the number of farms in each township of six square miles or 36 sections. These townships were grouped together with adjoining ones into large geographic areas of several thousand farms. These blocks of several hundred square miles became the AC price territories. They usually were bordered by a 25 kV line, a provincial border or geographic barrier like the Saskatchewan River, Last Mountain Lake or the Qu'Appelle Valley. From the number of farmsteads in that territory the amount of line needed to connect 80 per cent of them was calculated. This determined the cost each farmer would pay to get electricity. The farming areas of the north-east Parkland Belt with its more dense, smaller acreage mixed farms needed less line per customer than the larger acreage and more widely dispersed farms and ranches of the south-west. The AC prices varied from a low of $492 in the dense areas to a maximum of $600 in the sparse districts. The latter number was arbitrarily established as the most any farmer would pay to be served.

From experience we knew fewer farms in the more heavily settled parts of the country would electrify because of their economic health as they had smaller acreages with less income. Conversely, in areas with larger farms more of them would take electricity because of their better prosperity. In order to keep SPC's farm electrification subsidy as fair and equitable as possible for each farm connected, regardless of where they lived, a sign-up formula was developed to meet this goal. To achieve this a percentage of farms in each project would be required to take service. This formula required a greater percentage sign-up in the more sparsely settled areas and a smaller percentage in those more heavily populated. This was the sign-up percentage formula:

Number of Farms per Township	Sign-up Required
48 - 60	65 per cent
61 - 70	60 per cent
71 - 80	55 per cent
81 and over	50 per cent

* * *

TO ENCOURAGE the sign-up of new customers in each project the AC program was promoted with three marketing incentives. Firstly, in only the initial year in which a project was organized, the AC price was discounted by $50 to all applicants, a one time only "sale." Those who wanted service afterwards would pay the full AC price. Secondly, those farms already served within a project's boundary were included in the total count of farms and also credited to the number of required sign-ups. For example, if the total number of farms in a project was 100, and 65 per cent were needed, or 65, and if 10 already had electricity, then the sign-up target was 55 new applicants for that project. Thirdly, SPC helped to finance the AC construction cost, but there were limits to this loan assistance. One third of a project's applicants could make a down payment of one third of the AC cost with the balance to be repaid on their monthly power bill over three years with interest at five per cent. Two thirds of the farms on each project would therefore be required to pay the full AC price. It should be noted this condition was later relaxed and loans were given to two thirds of those applying with only a $100 down payment. SPC never checked on the credit rating of any farmer who applied for a loan, but this was never mentioned to any farmer or at any meeting. Their loans were always accepted. When farmers defaulted on loan payments the threat of removing the electrical facilities usually brought them in line.

There were only two AC projects organized and built in 1951 but that number mushroomed thereafter. This plan, over the several years of its life until the province was covered, connected the vast majority of Saskatchewan farms. There were a few jobs in the most remote parts of the south-west where AC simply did not work, as the prospective customers were so few and far between. Here, farms and ranches were connected on a mileage basis, the lines often travelling as the crow flies over the grasslands, direct from one set of buildings to the next, to keep SPC's investment lower and, more importantly, to give these people the same benefit as their provincial counterparts elsewhere.

The Farm Electrification Branch

THE FARM ELECTRIFICATION BRANCH under Charlie Smith's direction was one of several organizational units in the the Corporation's Sales Division, whose manager was Roy Sarsfield. He was a member of SPC's senior management, reported to General Manager Jack Tomlinson and, on occasion, acted as GM when Tomlinson was absent. The farm branch, with the largest staff of any of the other units, was responsible for planning and delivering the province's massive farm electrification program which was, without question, the major activity of the division and indeed, the Corporation.

As the mapping of farm sites by the municipalities was nearing completion in 1952 the populated southern half of the province was divided into three farm electrification regions and named the Regina, Saskatoon and Yorkton Areas, respectively. Each contained the same number of farms but the preponderance of smaller acreage farms in the north-east as compared to the rest of the province meant the Yorkton Area was the smallest of the three, geographically. It was also more diverse from a nationalistic perspective as colonization and immigration had attracted large numbers of farm folks from Europe and eastern Asia who brought along their rich ethnic culture, religion and language.

The boundary of the Yorkton Area was the Manitoba border on the east, south to Maryfield and Moose Mountain Provincial Park, then its perimeter angled north-west to Fort Qu'Appelle and on to Strasbourg and Last Mountain Lake, then northward to Lanigan, Humboldt, Weldon and Weirdale, east of Prince Albert. The Regina Area contained the balance of the south-east corner of the province along the Manitoba border, then westward following Saskatchewan's entire border with the USA to Alberta, north to the South Saskatchewan River, then easterly to Elbow, Craik and Fort Qu'Appelle. The Saskatoon Area contained the balance of the province to the north and west.

The Farm Electrification Branch office was located next door to SPC's Head Office, in what had been the Firestone Tire Store at 1753 Cornwall Street, space that was acquired to relieve overcrowding in the main building. Its storefront was an all-glass showroom built for automotive tire and battery displays. It now exhibited an array of electrical appliances: ranges, washers, clothes driers, deep-freezers and water heaters. One of Cass-Beggs's engineering reports about building and operating a provincial power system stressed the need to promote increased energy use after new customers were connected, to maximize the economic performance of the installed assets. This public display of "white goods" on loan from electrical equipment manufacturers and local appliance dealers was part of this promotional effort.

Reporting to Charlie were two field reps, Jim Rutherford and Dan Dojack, an engineering group headed by Grant Peterson, special assignments were handled by Murray Leach and a Chief Clerk, Bruce Reed. Me and two other farm electrification clerks reported to Reed. We each were assigned one of the three areas: Gord Seel dealt with Regina, Joe Licker had Saskatoon and I had the Yorkton Area. Assisting us were three junior clerks, all girls. That fall a new Yorkton Area field rep position was filled by H. Garnet Parcher. Dojack was then assigned the Saskatoon Area and Rutherford to Regina. Similarly, the engineering work was reallocated and new engineers hired, supported with technologists, material estimators and draughtsmen. Electrical engineer Frank Crandell was hired and assigned to Yorkton.

During the hiring process for new engineers Charlie met with Rutherford and I, the only volunteer Second World War veterans in the branch, and asked, "Do either of you object to our hiring an electrical engineer who was in the German army during the war?" Rutherford, a Canadian Army infantryman, had survived the tough North African, Sicilian and Italian campaigns. I had been in the Navy on the North Atlantic. Almost in unison we responded, "No, we're not about to refight the war. We've no grudges and our respective pasts should not pose any difficulty." The engineer was Siggy Bock, and allegedly had been conscripted to serve on the Russian front. There was greater potential for difficulty for Rutherford with Bock than for me as he would be the Regina Area engineer and they would work closely together. Resentments or problems arising from their opposite-side wartime service never arose to my knowledge, but Bock's arrogant personality was quite another matter.

All of us in the office were young rangy-tangs, enthusiastic, risk-takers and felt we could tackle anything new. We often didn't know how hard the tasks might be to accomplish. Failure was never an option. In fact, its possibility never entered our heads. This fervour and our relative youthfulness was noted by a man in charge of electrification of CPR facilities who referred to us as "Young whipper-snappers," because their offices were staffed by older people at the end of their careers who had earned their seniority on the rails. As our lines radiated out to all corners the province he was frequently in our office about service to hundreds of their various facilities: railway stations, freight sheds, steam engine round-houses, water towers, section houses and railway road crossing signals.

* * *

PLANS TO GET the 1953 farm electrification program underway shifted into high gear in August 1952, as did campaigning by the province's political parties for an election. A major platform plank in the CCF bid for a third consecutive term was the electrification of 40,000 farms and all incorporated towns by the end of 1956. While the 1952 program had connected an impressive 4,000 farms under AC, the party seeking reelection had presented an ominous challenge. What lay ahead was described in one of Charlie's memos as ". . . a relatively more ambitious program than has ever been carried out on this continent, or perhaps in the entire world!" The politicians did not make clear, as they are wont to do with election campaign promises, whether the 40,000 electrified farms were to be new connections or the total number served at the end of that four-year term. It was soon clarified the latter was our goal: 27,000 new farm customers, as 13,000 already enjoyed the benefits of electricity. Rhetoric aside, this was still a huge challenge as more power line would be built in the following four years than existed in the province at that time.

The CCF won reelection handily with 42 seats, the Liberals 11, a clear mandate to get on with their promised platform, including farm electrification. This was no political platitude, no idle promise that would be unkept as is the bent of many present-day politicians. While SPC's Board of Directors were charged by the government to meet the campaign promise, it was Charlie's Farm Branch staff, including me, who would, with ruthless determination and innovation, accomplish this formidable task over this vast province. It was clear we would be plowing new ground, not redigging familiar territory.

Charlie chaired weekly branch staff meetings in what we nicknamed, "The Bunker," an unpainted, concrete-walled, windowless, austere, basement storage room beneath the appliance showroom. We brain-stormed new ideas, debated policy and procedures which were an impediment to our work, pursued those which had promise, shot down those that didn't. Everyone was encouraged to contribute and participate. With innovation as the name of the game, The Bunker was a most unlikely place for planning and deciding the implementation plans for this important provincial program. We also received training there on new office procedures and forms, communication techniques, effective letter writing, how to handle combative farm delegations and irate individuals who lost their cool. We eagerly anticipated these sessions that were integral to the strategies needed to meet our targets.

We were a focussed, cohesive, no-nonsense, dedicated group, not only the implementation team, but the internal lobbyists that are so necessary to influence the changes needed to accomplish the mission we faced. They were exhilarating days!

* * *

IT WAS a hectic time, faced with the responsibilities for this major crusade with a new work force that needed training and, amid all of this, we were deluged with demands for service from an impatient clientele. Applications arrived by the basketful, individual farmers close to existing lines, but most were petitions from groups of farmers, in some case upwards of 100 names. A few petitions were on brown wrapping paper or on the back of a leftover roll of wallpaper, having originated at a meeting held on short notice in someone's living room, an elevator office, a curling rink waiting room, maybe even the beer parlour, where it was decided to get signatures while the iron was hot.

Rural neighbourhoods were encouraged to apply as a group, to complete petition forms we supplied showing the land locations of each signers' farm site and to appoint a pro-tem chairman. These figure-heads were always men, a biased sign of the times, usually community leaders, a school district chairman, an RM reeve or councillor, the president of the hospital board or a Saskatchewan Wheat Pool representative. Often the petitions were delivered to our office by up to 8 or 10 of their signatories, decked out for the trip to the city in their unaccustomed shoes, suits, shirts and ties, and would have preferred to be in their overalls and work boots more befitting the plow-jockeys they were. Their uncomfortable dress was far outweighed by their zeal, determination and sincerity about getting electricity.

When group applications were received either by mail, in person or by delegation each was given a name, descriptive of their location and identified by nearby communities, centres that would be served only if the farms were successful, like Hampton-Donwell, St. Front, Veregin, Kipling-Kennedy, West

Bend. Others were named by their proximity to the nearest community, such as Whitewood South, Choiceland North, Rose Valley East, Pleasantdale West. Once named it stuck with that group until they were served.

In attempts to pressure and influence us the delegations frequently dragged along their Member of the Legislative Assembly (MLA) or Member of Parliament (MP), or their RM reeve, or the mayor of a town or village overseer whose citizens also wanted electricity. A few had the wisdom to be escorted by an influential person knowledgeable about our program, like Father Matthew Michael of St. Benedict or Charlie Vickers, a Brooksby storekeeper, or George Lincoln from Wawota, the RM Reeve. These men knew the ropes of farm electrification and had proven records for helping their own and many other groups get service.

Wednesdays were our busiest days, when rural community stores closed for the afternoon. Regina businesses were happy to have the wives shop while their husbands lined up to see us. This pattern continued throughout the year except for a few days when farmers made hay, seeded or harvested. Surrounded by eager applicants, our program was explained with a provincial map and one of their locality draped over our counter, or more often than not, on top of a freezer or clothes drier. These maps showed the colour-coded four-year program along with locations of the thousands of petitioners like themselves and the tentative line locations that would connect them. They questioned and presented determined, even desperate, arguments as to why they should be considered ahead of others, particularly if their 32-volt windcharger and battery systems were on their last legs. We were on our feet most of every day and if there were heavy line-ups we were supported by one or more of the field reps, if they were in town, or sometimes by Charlie. The need to repeat our message to group after group was tiresome, compounded by uncomfortable heat generated by the sun on the south-facing windows in the non-air conditioned showroom.

We tried hard to make petitioners feel they were getting special attention and that we were not mindlessly dispensing the same story hour after hour, but we were. We ensured each delegation understood the enormity of our task and how they fit into the program. After all, good logic, sensible, well thought-out plans and thorough explanations are readily accepted by those who have a relationship with the land. Some groups were very vocal, and lobbied hard and frequently. Inevitably some would wait up to three years, maybe longer, before we could serve them because of their proximity to existing power lines. It was awkward to have to deliver this bad news as they found it frustrating and tempers did flare at times but this was more the exception than the rule. Every enquiry we dealt with was noted and commented on the next time we met anyone connected to that application. We wanted farmers to know we had farm electrification well under control.

Most politicians accepted our four-year plan but challenges were often made directly to SPC's Board of Directors Chairman, Jim Darling, in an attempt to exert some political clout. He understood and was sympathetic to our workloads, even instructed his CCF caucus colleagues to avoid putting us under undue pressure. But there were exceptions. The two worst, by far, were Jim Thair, MLA for the Lumsden constituency in the provincial legislature, and W. Ross Thatcher, MP for the Moose Jaw-Thunder Creek federal constituency. Both flew the CCF banner and one would have thought would have been more understanding and sympathetic, but they were obstreperous and overdosed with self-importance. Thair's demeanor emulated an irritating, pesky fox terrier. Thatcher, known to abuse alcohol, was a mean man, impatient and pompous, often impolite and impertinent, full of bravado, who taught his murderous son, Colin, all he knew.

Thatcher always wore a dark blue suit and a matching rolled brimmed hat, a stick pin in his tie and, in cooler weather, a dark topcoat, chewing on a fat cigar. Each time we met delegations with these two politicians they impatiently listened, already familiar with our story and then, when we had finished our spiel they would predictably ask, "We want to see your supervisor," as if to say, we don't like your story, we'd like a better one from your boss. They both had full-grown, developed egos and prejudices inherited without much thought. Their show-off deportment reminded me of the comment about elephants copulating – it took place at a high level accompanied by much noise and mountainous effort. They and their followers were referred to a more senior person, a field rep, maybe Charlie, sometimes Sarsfield, who repeated our story. While there was no doubt they had their constituents' best interests at heart this attitude did not endear them to us nor to the interests of their constituents, but these groups did not suffer delays of any kind in getting service because of the personal traits of the politicos who accompanied them, nor did they get preferential treatment over others. We were never overruled either within SPC or by the CCF government, to their credit, to prioritize one group over another. This is not to say there wasn't pressure that was heavy at times, but once our program plans were made and publicly announced, we ably withstood the most intense lobbies.

In our discussion with one delegation with our maps on top of a deep freeze, another group listened intently as they awaited their turn. And often we had other listeners too: General Manager Jack Tomlinson, when he entered or left his adjoining office, Darling or other board members entering or leaving the nearby boardroom, and SPC executives as they passed through the display area to and from their offices. They often stopped, listened and sometimes commented to reinforce what we were telling the farmers. This support was appreciated as we were on the front line.

While most farm groups came from the Regina Area, because of Head Office's proximity to them, the delegations were by no means restricted to just that area. Each clerk dealt with his area's enquiries, but no group waited for the appropriate clerk. Although not required of them, the girls pitched in too. Counter demands often continued uninterrupted for days. It was difficult to maintain office procedures, correspondence and administrative work, so Charlie often recruited Leach and Reed, and all three would join the three of us for several hours for a night of overtime to catch up.

A problem faced with office delegations and also at organizational meetings by those of us working in the Yorkton Area came from farmers living close to the Manitoba border. There, the provision of electricity came at no cost to farmers under the Manitoba Hydro program. As the Yorkton Area's eastern boundary was the Manitoba border, this justification to Saskatchewan farmers to pay some of the costs while their neighbours across the road in Manitoba paid nothing was a continuing one. Conversely, on the province's western border, adjacent to Alberta, the cost to get electricity was seen by Wheat Province farmers living there as a bargain when compared to all the power line costs charged each farm by the privately owned utilities, Alberta Power and Canadian Utilities.

* * *

WHEN DISCUSSIONS in The Bunker led to suggested changes in a procedure or condition of service, or a new policy was needed which was outside Charlie's authority, approval was sought from Sarsfield or Tomlinson and sometimes a Board of Directors decision was required. Occasionally, legislative amendments to the REA were required. The frequency, importance and timeliness of our requests reached the stage where the Board of Directors appointed three of its members to a Farm Electrification Sub-Committee, with decision-making authority. They were Irwin Hockley of Indian Head, Tom Lax of Regina and Harry Marsh from Herschel, the sub-committee's chairman. All were retired farmers, staunch CCFers and had been rural community leaders who readily understood the issues we faced. Sometimes Charlie had us join him to plead our case. It was a privilege to become acquainted with these men, an opportunity we three unionized clerks had that was not afforded other Corporation employees.

In 1953 the phenomenon of farm family migration to urban centres started to give us sign-up difficulties. This abandonment of habitable farm sites presented us with projects where every set of occupied buildings, including those who rented the land and lived on the property, had signed for service but the group was still short of the target number of needed sign-ups. It was sheer nonsense to insist on more applicants when there weren't any or to demand that we connect our lines to vacant buildings. To turn down a project and refund money already collected was unproductive. Doing so meant organizing a replacement group someplace else with no guarantee of its success. Those we turned away would be denied service, perhaps for a couple of years, a dilemma they would not accept gracefully and, more than likely, would raise in the murky political arena.

I was asked by Charlie to present to the board's Farm Electrification Sub-Committee a case study on the vacant farm issue which we had encountered on our "Finnie North" project. I was nervous as I made my submission, trying hard to be convincing so as to garner their approval. I concluded saying, "It was very difficult to work closely with the local farm committee, to not be swept-up in their zeal and be tempted to bend our regulations, particularly when they seemed unproductive, illogical and didn't make sense." Harry Marsh, a very tall, imposing man with an even bigger voice, mindful of the issue and also the importance of following the rules, responded with a twinkle in his eye, "Young man, we understand you want to help these farmers but, don't forget from where your pay-cheque comes." The sub-committee, without hesitation, agreed with the recommendation to reduce the total number of farms in a project by up to 10 per cent where vacant farm sites presented this problem. This relief, where it was needed, was a most helpful change

* * *

MY FIRST ENCOUNTER with a farm project committee on their territory was on the "Dafoe South" job in 1952 with Garnet Parcher, just after his appointment and when the farmers had finished harvesting. Their sign-up work was complete. This large project had been organized by Jim Rutherford that spring. Crops in this light soil area had been above average that year so there was no shortage of money or the desire for electricity. Our target was about 90 farms, but well over 100 had signed, a remarkable achievement in this predominantly Scandinavian district. It was an unreal baptism for Parcher and me. We commented on the drive back home, "If all projects are this easy, the task ahead does not seem to be as formidable as we've been led to believe." We soon learned that "Dafoe South" was an exception to what we would soon experience.

Another group that attained their sign-up with ease that year was "Abernethy North," a prosperous farming district. While not supporters of the sitting government by a long shot they had some difficulty believing they would get service because of their political leanings. Overcoming their reticence gave me a good feeling, able to prove that our purpose was to serve all farmers, not just those who supported the sitting government. Their project included the hamlet of Gillespie, named after a district pioneer, a lone Saskatchewan Wheat Pool elevator on the railway line between Balcarres and Melville. We electrified it, but this prairie sentinel, and thousands of others of these distinguishing landmarks, have since vanished from the skyline.

New power lines served both of these projects in 1953.

* * *

TO HELP ME understand the relationship between our farm sign-up work and that of commercial field reps, who handled the electrification details with residents in new communities that were about to be served, I accompanied Mel Martin in the spring of 1953 to the hamlet of Serath, midway between Raymore and Southey on #6 Highway, north of Regina. Serath was an inland community in that it was not on a railway line. This hamlet would get electricity because our "Raymore South" project committee had attained their sign-up under the able chairmanship of Julius Orthner, a quiet spoken farm community leader. The townsfolk had earlier been notified of the costs, something around $50 each, much less than what farmers paid. Martin held a short meeting and the people were given assurances the lines would be built. Each customer signed a contract and we wrote receipts as they paid their money. The secretary of the RM, administrator of the hamlet, signed the street light contract as they would pay that account. This community sign-up work was much less onerous and uncomplicated than the struggle the farm reps had in their assignments.

About 15 years later one afternoon at Regina's Exhibition, lined up at an amusement ride with my children in tow, a farmer and I eyed one another, searching to put names to our faces. Recognition happened quickly. It was Julius Orthner. We warmly greeted each other and chatted briefly. He still spoke warmly of our work to help him and his neighbours get electricity.

The Farm Electrification Program

BY THE SUMMER of 1952 we had received additional information about the rural community, data that was added to our maps of farm site locations and AC pricing regions. The Department of Municipal Affairs supplied land assessment values for each township, the basis for setting mill rates and raising taxes by the RMs. This identified areas with the most productive land and the better buildings, probably occupied by more prosperous farmers better able to pay the AC construction charges, wire their buildings and buy and operate electrical equipment. This information also identified submarginal acreages where eking out a living from the land was a huge challenge for folks who might not be able to be our future customers. As well, the Department of Agriculture supplied the average wheat yields and cattle population for each municipality. This pinpointed mixed or dairy farming districts, where electricity usage would be higher for handling supplies of feed and water, and in dairies for milking machines and refrigeration. Our services would be more beneficial to these users than those regions best suited to only cereal grain production.

We went one step further. The farm reps met with Department of Agriculture field representatives (ag reps.) They knew the farming districts intimately because of educational and community betterment programs and activities they conducted like soil analyses for more efficient crop selection, fertilizer use, farm management practices, livestock breeding, the proper selection, care and use of farm equipment and 4H programs with farm youth. The ag reps could rate the farming districts from their perspective, determine which were the most innovative, and maximized the conditions under which they farmed. Their ratings ranged from excellent, through good and fair, to poor. While not hard statistical data, it helped us decide which areas we should electrify ahead of others when there was an option. This ag rep information was given to us with the understanding it would be held in strict confidence.

An override to all this data and opinions was local farm interest and the order in which groups had petitioned us for service. It was unfair and impossible to defend if one group was served ahead of one or others that had applied long before, assuming they were of similar status with all the other information we had gathered.

After consulting with the power line construction people about their gearing up with men and machinery needed for the job ahead, we decided the four-year program targets would see 5,500 new farms connected in 1953, 6,500 in 1954 and then expand to 7,500 in each of 1955 and 1956. In addition were the many thousands of citizens living in communities that would be served coincidentally. The targeted 27,000 farms were equally divided at 9,000 for each of the three areas. As it required an average of one and one tenth of a mile of new line per farm, this goal meant that 30,000 miles of 13.8 kV single-wire farm lines would be built in those four years, needing a whopping 425,000 power poles. In addition would be the hundreds of miles of new three-wire 25 kV distribution lines plus the poles, conductor, hardware, transformers, meters and street lights needed for the thousands of customers in the urban centres that would be served as the farm program unfolded.

We encountered difficulties in some parts of the province where there was an absence of 25 kV distribution lines and the higher voltage 69 kV transmission lines, both needed as sources of supply to the single wire farm lines. The most notable locales were in the ranch lands of the south-west and in the north-west from North Battleford to Meadow Lake and west to Pierceland. So our farm electrification plans were coordinated with SPC's transmission line and substation planning groups and their construction schedules. At times the rural electrification program hastened the building of these other facilities.

It seemed inevitable the more vocal and well organized a farm group was the further it was away from the nearest source of power. This meant to reach those with the strongest voices and interest the lines needed to be built through other farming communities in the early years of the program. A case in point was a large group of good, mixed farms west of Wawota. They would be supplied from the 25 kV line which ended at Kipling, over 30 miles away. Before serving them a project from Kipling to Kennedy had to be organized, signed-up and built. Then, in the following year, or maybe even two, the Wawota group could be picked up. This was hard for them to accept as the Kennedy farmers were not as well prepared and with no guarantee that next year's crop and farm markets would assure the Kennedy job, it left many uncertainties the Wawota group had to live with.

We three farm clerks intimately knew the whereabouts of the interested groups in our respective areas, as much or more than the field reps. We huddled for days with them and the engineers to lay out the 1953 to 1956 program: 27,000 new farms, our targeted number. Project sizes varied from 20 new signers to over 100, about 40 or so being the average. Sometimes a project's size included more than one group of petitioners.

The detailed four-year program, project by project, to meet the election promise was needed for three reasons. Firstly, approval of the plan by the Provincial Cabinet was required as they would borrow the necessary funds from the money market brokers and investment houses on Bay Street in Toronto, Montreal's James Street and Wall Street in New York. Secondly, an orderly plan enabled the farm community to know when they could expect service, important for owners of 32-volt windcharger systems near the end of their useful life which could perhaps be coaxed along for another year or two until power arrived. Thirdly, the plan was needed to assuage the demands of thousands of people in the small urban communities who were also clambering for service.

All the farm data we had gathered, except the ag rep information, was displayed on a provincial map, about four feet square, prepared by Murray Leach. Those projects in each of the four years were colour coded on overlay plastic sheets, one on top of the other. This impressively demonstrated how much of the province would be covered by the end of 1956, the end of the government commitment. Area coverage was indeed an appropriate name. Additional overlays showed the land assessments, cattle population and wheat yields. An extra overlay, intended only for our internal use, was added to show the provincial electoral constituencies, so no MLA, government or opposition, could claim unequal or unfair treatment of their constituents. It was an impressive presentation.

The charts, maps and the capital costs for the four-year program were presented to SPC management, the Farm Electrification Sub-Committee, then the full Board of Directors, who in turn recommended it be shown to the premier and Cabinet.

When Charlie, Tomlinson and Sarsfield returned after their presentation to Premier Douglas's Cabinet we were called into Sarsfield's office to hear the outcome. It had been accepted and the Cabinet, to their credit, made no attempt to reassign more to CCF constituencies, at the expense of the Liberal opposition members. The Cabinet had gone one step further and asked for the plan to be presented to all MLAs, wanting to ensure there was understanding and support from all the elected members. Roy Sarsfield made this presentation a few days later in his usual confident and flamboyant style. As Sarsfield concluded an Opposition member asked, "How much did politics enter into your selection of the projects to be electrified?" Sarsfield emphatically replied, "None," just as he flipped the last overlay that ended his visual presentation. It was the provincial constituency overlay, brazenly obvious for all to see. It neglectfully had not been removed beforehand. With his credibility wounded, we had done our homework as the map clearly showed an absence of political favouritism. The Opposition had little to complain about and gave their support.

With the political approval in our hip pocket we were now ready to shift into high gear. In fact, we went into overdrive. Immediately the committee chairman of each farm group was contacted about holding organizational meetings to get the sign-ups underway for the 1953 program. The commercial field reps contacted the town clerks, village overseers and municipal secretaries with the news about when their respective communities would get service. Farm groups we expected to serve in 1954 and succeeding years were similarly advised of our plans.

A Little Bit of Gas – Naturally

A BY-PRODUCT of oil exploration in western Saskatchewan was the discovery of natural gas. In 1951 the government designated SPC as the sole purchaser and distributor of this natural resource. It was initially intended as a replacement for diesel fuel burned in the engines that produced electricity in power stations for two important reasons, as it was cheaper and burned cleaner. In 1952 the diesel engines in the Kindersley Power Station were converted to burn natural gas and, as well, 290 residences, stores and businesses in Kindersley and Brock were served.

In 1952, David Cass-Beggs, the government's energy consultant on electrical matters, was asked for advice on the merits and problems of establishing a provincial Natural Gas System. His *Report on the Economics of Transmission and Distribution of Natural Gas in the Province of Saskatchewan* became the basis for SPC to launch itself into the gas business. When adequate gas supplies and long-time reserves were proven-up that year, planning started to bring this new heating fuel to the city of Saskatoon the following year.

This birthed SPC's role as a dual utility. Those in the energy industry referred to monoliths like ours as Roman Riders, charging ahead with one foot on each of two galloping horses. Cass-Beggs believed setting up a second crown corporation did not make sense considering that many internal functions needed for SPC's electric system were already in place and could readily be used by the natural gas system – personnel, labour relations, legal, meter reading, billing of accounts, public relations, accounting, vehicles and equipment, purchasing, office services. The government bought the concept. John Mollard, manager of the diesel power stations in the western part of the province became Gas Development Engineer and later the Natural Gas System Manager. To technically guide the Corporation in this new venture, an experienced American natural gas consultant was engaged: the Fish Engineering Company of Houston, Texas.

Saskatoon's supply would come from the Brock natural gas field through a 120 mile, ten-inch, high pressure, welded steel pipeline. This would be a North American "first" where a pipeline of significant length and a distribution system for a medium sized city would be built and in operation in the same year. Not only would the city be served, but the tiny hamlets of D'Arcy, Fiske and McGee would be connected at the same time as the line went right past their door.

In the spring of 1953 a natural gas office was opened in Saskatoon, supervised by Mel Hunter, a Regina based commercial sales rep. Mel was a free-wheeler, not endeared to auditors, rules and routines. His assignment was to conduct a sign-up campaign with customers to have gas service pipes put into their premises and to assist the Department of Labour's Gas Inspections Branch to train and license heating and plumbing contractors who would be needed to install gas lines and equipment inside the homes and businesses in time for them to use this new fuel the coming winter.

In July, after the office had been open for a couple of months Roy Sarsfield asked me to go to the Bridge City to organize that office, in spite of the heavy farm electrification demands in which I was embroiled. They needed better office routines, a more secure collection and banking records system for customer service deposits and coordination of service applications with the daily construction activity of laying the gas lines through the city's streets and back lanes. I readily agreed to go as it was a new challenge for me, but I was surprised and bewildered - why me? Some years later I learned the extent of Sarsfield's special interest in me. More will be said of this later.

I stayed in Saskatoon's then rather palatial King George Hotel. One weekend my wife came, somehow at company expense, giving her a short break from caring for our one-year-old son. No doubt Mel Hunter was involved in this benefit for us.

One Monday morning I hitched a ride to Saskatoon with General Manager Jack Tomlinson, and this was memorable for three reasons. First, as a unionized clerk I became acquainted with the big boss. Although small in stature, I found him to be a bit of a cavalier, dominating and undiplomatic. Second, his car, a company-owned Mercury sedan, was luxurious as compared to the lowest-tendered-price company pool vehicles in which I usually rode. Thirdly, he drove like the wind, up to 80 miles per hour, even on the gravelled #11 Highway between the two cities, calmly making roll-your-owns as he steered with his elbows.

The office was a bit of a mess and I worked a lot of overtime during the three or four weeks I was there, so it was a lucrative month for me, money-wise. Hunter had hired a retired bank clerk, Jack Evans, to run the office, but he was unfamiliar with Corporation routines so it was necessary for me to train him. We established record systems and procedures, which pleased SPC's treasurer Erwin Stuhr, Hunter and the pipeline contractors. By year-end over 7,000 customers had applied for and had service pipes run to their buildings from the city's back lane natural gas mains, more than 10 per cent over the target, of whom 4,700 became active users.

I watched natural gas arrive in Saskatoon for the first time when the torpedo-like "pig" arrived at the regulator station on the city's outskirts, pushed through the pipeline by the gas pressure to clean out dirt and construction debris before it went into operation. I observed the various construction phases on the streets and bridges, and in the lanes and back yards. My stint in Saskatoon had been an interesting experience but I was glad when it was over and I could get back to what I thought was more important work.

To Yorkton

PRIOR TO MY Saskatoon natural gas sojourn it had been decided head office's Rural Electrification Branch would be decentralized. The Regina Area staff would stay put but the other two areas would open offices in Saskatoon and Yorkton, respectively, closer to their farm clientele.

This move was the subject at several Bunker sessions that raised a host of questions: who would go to which area; what date would the decentralization happen; would SPC pay the costs to move our families: would SPC pay for travel and subsistence expenses until we could find places to live; where would the new offices be located; if Regina based employees did not agree to relocate new staff would have to be hired and trained; would office supplies, furniture and equipment be shipped from Regina; who would make banking arrangements for the money we collected; advertisements and publicity was needed to inform the rural population in the areas about the effective date of the moves, the new office addresses and telephone numbers.

Charlie asked for volunteers for the Yorkton and Saskatoon offices. I was content with the former. I was, by then, familiar with the area, made easier by my travelling the same territory in 1948 for Steele Briggs. More importantly, my new supervisor would be Garnet Parcher. He and I were friendly and worked well together. My wife, infant son and I needed new accommodations even if we had stayed in Regina, as we had occupied her mother's home for a year while she visited Australian relatives. The moving decision was easy for us.

From my exposure to farmers at Steele Briggs and in the Farm Branch, I liked working with them, knowing sod-busters are impossible to deceive on what is good for them and their farm. A doctor, an engineer, an artist, a lawyer, a soldier, can all impersonate and be impersonated for a short while with some hope of success. A farmer never. These men of the prairie have been conditioned to the pioneer ways wherein individual initiative and innovation were commonplace necessities. There is something mystical about a man of the soil when he picks up a handful, lets it run through his fingers and wonders about the mystery of life it embodies.

My study at university of Stuart Chase's essay, "*The Luxury of Integrity*," indelibly imprinted its contents on me. As an occupation, when compared to other professions, Chase wrote, "The independent farmer, standing closest to the pioneer tradition, leads the list – for agriculture is far more a way of life than a pecuniary pursuit." For a farmer all the security and wealth in the world does not look so prosperous nor so safe as a flock, or a herd, or a field of ripe grain, or a summer-fallowed field. Flocks and herds, land and crops are real. They appeal to instincts which are as old as man, and this attracted me. The Yorkton move was part of this appeal as my work was self-satisfying and the farm electrification job ahead was a massive challenge. I believed we were at the forefront of an historical venture, making significant contributions to improve the lives of rural people.

Amongst the Regina staff assigned to the Yorkton Area work, only engineer Frank Crandell and technologist, Jim Pirsak, agreed to relocate. Pirsak was single, but Crandell's young family, like mine, needed a new place to live even if they had stayed in Regina. My clerical assistant, a married lady, could not move so we would need new staff for the positions of junior clerk, secretary, draughtsman and estimator. These unionized jobs were advertised internally before the move took place but there were no applicants. We would have to hire these people after we had moved, so 50 per cent of our new office work force would be inexperienced, which added more pressure to our work.

* * *

CHARLIE TOOK ME to Yorkton in August to scout for available office space. As well, he wanted to observe Parcher conducting an organizational meeting, as he was new to the field rep position. Vacant office space was nowhere to be found. The city had grown rapidly since the end of World War II. New commercial building construction had not kept pace with population growth and not only was there an absence of office accommodation, but also an acute shortage of rentable homes, suites and apartments. Finding living space for our families would have to be left until after the office relocation when we were closer to the situation.

The only option for office space we could find was the unused basement of SPC's Yorkton District Electrical Superintendency headquarters on Third Avenue. It was rented space, owned by the weekly newspaper, *The Yorkton Enterprise*. A.E. (Bert) Allan, District Superintendent, along with George Sinclair, District Clerk, and two cashier/stenographers, Mabel Anderson and Sheila Linfoot, occupied the main floor. We arranged to have the basement cleaned out, the concrete ceilings and walls painted white and some wiring done for outlets and fluorescent lighting. Our farm customers would have to traipse through the District Office, past Allen's office door, and descend the stairs at the rear of the main floor into our den. This windowless space was barely seven feet high with no ventilation, made worse by several of us who smoked, as did a majority of our clients.

When deciding on this space we were unaware that every Thursday was newspaper printing day and from early morning until mid-afternoon the presses ran. They were in the basement adjoining us and the noise meant that any discussion between staff or with those who came to do business with us was only possible by shouting. In addition to this disadvantage was the objectionable smell of printers ink that permeated our space, and the heat generated by the motors and machinery, tolerable on colder days but uncomfortable on warm ones. Present day labour unions, including ours, by the late 50s, would not have allowed their members to work in such dreadful conditions. We existed there for over a year, overlooking the office's shortcomings because of the daily challenges that engulfed us as we laboured to meet our targets of signed up farmers.

* * *

PARCHER CONDUCTED the two organizational meetings with Charlie and I sitting at the back of the classrooms. The first was at Calder, east of Yorkton. This project abutted the Saskatchewan-Manitoba border, its farm group chairman was Steve Surjik, an astute man and community leader who effectively managed his sign-up committee as they handily met their target. Surjik later moved into Yorkton where we became better acquainted. The second meeting was at Ebenezer, 10 miles north of the city. Garnet did fine at each session, albeit a bit queasily with his boss and me in the back row. On the day of each meeting the three of us field-checked the project by car to verify the RM mapping data and farm site locations.

Rain on the first day at Calder made many roads impassable and put a crimp in our task but the second was sunny and warm. Charlie had his .22 calibre revolver and from the passenger side attempted to shoot some ducks as we came across roadside sloughs. He hoped they would be our dinner but his aim and best intentions were unsuccessful.

At one point when I was driving he asked me to stop suddenly when he spotted wild mushrooms growing along the edge of the road. While his lips smacked in anticipation of this gastronomic treat his pocket knife carefully cut them from below ground level and then he gently placed them in his overturned, gray fedora. Meanwhile, Parcher and I kicked at them nonchalantly in the gravel, much to his scorn. He scolded us and said, "C'mon you two, help me pick these up. They'll be our supper instead of the mallards." With little enthusiasm we helped fill his hat to overflowing, then we also filled an empty beer carton found in the ditch. We vowed Charlie would dine alone.

With our field check finished we returned late that afternoon for dinner in the small cafe in Ebenezer that had been our lunch venue. Its owners, a young couple, were about to close and leave for Yorkton where their days-old premature baby was still hospitalized. They sort of knew us from lunch and said they would be gone about an hour. Charlie, carrying the box of mushrooms, persuaded them to let us use their kitchen while they were away, a surprising gesture of trust. After they left he played chef, and washed, sliced and cooked our afternoon harvest. Soon he served us plates piled high with butter-fried, blackened mushrooms, bacon and toast. Parcher and I conceded if he ate them we supposed his desire to live was as great as ours, so we did too. Admittedly, they were delicious. We paid the owners for the food and our use of their kitchen on their return and went to our meeting.

On the highway close to midnight as we headed towards Yorkton, Charlie asked, "You guys feel okay?" Assured that we were, he then admitted to his study of prairie mushrooms and his knowledge of which were edible and which weren't. He loaned me a book on the subject, and in later years after a summer rain my children and I often picked them on rural roads when at our cottage on Echo Lake in the Qu'Appelle Valley.

* * *

ON THE SEPTEMBER 1953 Labour Day weekend, moving vans took our office furniture, files, maps and supplies to each of Saskatoon and Yorkton in time for business on Tuesday morning. The move was timed to coincide with harvesting that was in full swing in the farming community, when clients were too busy to come to our office unless it rained. Crandell, Pirsak and I drove in my car, a 1947 Chevrolet sedan, to Yorkton on the holiday Monday evening, the first of many commutes. Even though we had volunteered to move, which was of benefit to the Corporation, the staff expense allowance limits for unionized employees prohibited our occupancy of West Broadway's new Shamrock Motel, luxurious as compared to the old and austere

Yorkton Hotel, with its bathroom down the hall. We lived there five nights a week for three months. Staying in the more luxurious motel would have cost about $2 more per night for each of us. We made the return trip to our homes in Regina on Friday evenings right after work, anxious to be with our families. These prolonged absences were an emotional strain.

Hotel life and restaurant food day after day was mundane, but one night newlyweds occupied the room across the hall from mine. It being a warm evening, the transoms above our respective doors were open for ventilation. It was impossible to not overhear the couple, alone at last, as they prepared to consummate their marriage – he an excited and naive groom, she an experienced and impatient bride.

Not surprisingly from my earlier scouting expedition with Charlie for office space, Crandell and I had difficulty finding accommodations for our families. My wife and I finally rented a 70-year-old house that had been one of Yorkton's first churches, built in 1880, at 67 Tupper Avenue, for $30 a month. The owner's unethical rental agent, a Mr. Thompson, asked for "a-little-something-on-the-side," $20 under the table, just to ensure we got the place. This self-anointed pew-sitter professed his Christian ethics on Sunday mornings in the same church we attended. In late November we moved and existed for 13 months in a house void of storm windows or of insulation, inadequately wired and next to impossible to heat for the comfort of my wife and our 17-month-old son.

I received the car mileage allowance for the first month, after which the company comptroller, Frank Copithorne, refused to honour my claim, contending it was contrary to provisions of the union agreement which stipulated only 30 days of travel expense. Before the move Charlie believed, but had failed to confirm, that the Corporation would continue the arrangement because we had volunteered to move, but such was not to be the case. He tried, unsuccessfully, to intervene and have the decision overturned as we three rode together and there was only one claim for the car allowance. As well, the travel time to and from our job location was on our own time and on some Friday and

Sunday nights we worked with project committees that were out of our way and off the highway. The outcome of this conundrum was that each of us would be reimbursed for travel only to the amount of equivalent bus fare. SPC's saving were but a pittance. We continued to use my car but to offset my vehicle costs Crandell and Pirsak gave me the bus fares they claimed. This was a chintzy deal imposed by the comptroller, who had little experience of travelling life and rarely left his office, comfortably at home with his family every night. I had expected a more reasoned and conscionable approach from Copithorne, whom I had known when since starting at SPC in the accounting job.

His act of saving money, no matter what it cost, bemused me at first. Then, the more I thought about it the more I was ticked-off and angry at this unfair and indecent treatment. However, because of responsibilities to my wife and son and without having been born with a silver spoon in my mouth or any other back-up resources for their welfare except my monthly pay-cheque, the challenges of the job and the excitement of the move of both office and home derailed my instincts and common sense. So instead of making an issue out of this wrongful act, it was shuffled to the back of my mind. No doubt the Corporation was also aware of my situation and was exploiting it. It was a signal to which I should have given more import. They expected me to innovate and produce at every opportunity and yet they fell so far short when the other side of that coin was exposed.

* * *

The Quarter Section Farm[1]

I stopped a while on a lonely road
Where a farm yard used to be.
It now formed part of an endless field
Where a proud young farmer walked tall.

---

[1] From *Living Skies and Saskatoon Pies – Stories and Rhymes of bygone times* – by Claude Finch, a former SPC employee.

A man and his wife raised a family there
Until progress ruled the day.
Their hopes and dreams were shattered
As they sadly moved away.

I found the spot where the house once stood
Now there was nothing at all.
A place that once rang with laughter
Not even a shrub or a tree.

The only reminders of the past
Was a mound where the rhubarb grew,
A handle from a teacup
And the sole from a child's shoe.

* * *

ON ONE SUNDAY evening commute we stopped a few miles
off the highway at Eric Welsh's farm. He was the fiery, red-
headed, farm group chairman of the "Finnie North" project. His
committee had completed their sign-up and our visit was to
gather the completed forms and money. Three other canvassers
and their wives were already there when we arrived, apprehension
etched on their faces.

When we checked their paperwork they were two farms
short of the needed number of signers. They emphatically
pleaded their case that every farm where someone lived had
signed-up, including several renters. More applicants were just
not possible. Crandell, a bit of a fussbudget and a stickler for
detail, like most engineers, insisted on two more sign-ups. Trying
to deflect his obstinacy that was head-longing us toward the
brink of a touchy confrontation with the excitable Welch, I asked
why and how long the vacant but habitable sets of buildings had
been empty, and about the possibility of their future occupancy.
Most were suitcase farmers, coming in the morning to work their
farms and returning to town at night. I noted these on the project
map. It was apparent Crandell and our regulations were

unrealistic. It seemed blatantly silly, at best, to insist on electrifying empty farm yards.

Common sense got the better of me so I committed a serious review of their plight, thus overriding Crandell's objection. After lunch we left a disconsolate group. While well aware of this shift towards vacant farm sites this was our first case where every occupied farm had signed up and the project was still short the number we required. As mentioned earlier, this was the case I presented to the Farm Electrification Sub-Committee of the Board of Directors, which agreed to alter the policy to recognize this new societal trend, to reduce by up to 10 per cent the number of farms in a project where this condition existed.

"Finnie North" was served the following year with the number of farms Welsh and his committee had garnered.

* * *

THAT FIRST WEEK in Yorkton was a hurly-burly time as we located desks, filing cabinets, supply cupboards, drafting tables and client interview areas to suit our needs amid the under-foot installers of our telephone system. I arranged for a deposit and transfer account for the hundreds of thousands of dollars that would go through our hands each month with Dave Forsythe and Brian Campbell, the manager and his assistant, respectively, of the Royal Bank branch a few doors south of our office. They were delightfully surprised with the volume of unexpected business that fell into their laps as it raised their branch's status in the bank echelons. My business relationship with them was helpful in a personal way when, a year later, as I struggled to arrange financing to build a new home for my family, Forsythe, having come to know me, had his bank carry my mortgage.

We hired a stenographer, Barbara Hodson. She had just resigned from Bowman Bros., an automotive parts supplier, and planned to go to Saskatoon where a brother lived. Her father died unexpectedly after her resignation and before she moved. Feeling compelled to stay in Yorkton to support her mother, she needed

work. Crandell and I met with Russ Baldwin, principal of the
Yorkton Regional High School looking for recent grade 12
graduates who had good draughting and mathematical aptitudes.
On his recommendation we interviewed and hired two local boys,
Graham Hall for the draughtsman position and Atley Lovelace to
be the construction material and cost estimator. All three
continued working in the Farm Branch until the program started
to wind down some years later, then they continued with SPC in
other roles in other locations until their retirements.

* * *

THAT FALL, work demands to fulfill the commitments of the
farm electrification program required that another field rep be
hired in each of the three areas, so notices for the new positions
were posted. These jobs were considered to be management
roles, outside the jurisdiction of our trade union's collective
bargaining agreement, because the incumbents were expected to
work irregular and longer hours. This manoeuvre was a company
ploy to avoid having to pay the penalty of overtime costs had the
positions been within the scope of the union contact.

I bid on the Yorkton job but my seniority was insufficient
to land the assignment. It went to Harry Jessop, the Aneroid DO.
He was a tall man, over six and a half feet, and as thin as a rail.
He could not stand upright in our basement office. Jessop was
known amongst his peers as the step ladder district operator as
he could not, for some reason, climb wood poles with a belt and
spurs as was commonplace for other linemen. I did not take to
him at all, but for reasons that will be told later.

* * *

SAM WYNN WAS publisher of *The Yorkton Enterprise*
weekly newspaper. He was an aloof, small, quiet-spoken man
who cherished his University of Saskatchewan honourary
doctorate, and liked to be referred to as Dr. Wynn. He was a
respected elder statesman of the Saskatchewan Weekly

Newspaper Association.

The day of our move into the Yorkton district office we happened to meet at the front door just as he arrived for his day's work. He was curious about the moving van presence and we talked briefly about why there would be more SPC staff in the building. I promised to explain our operation after we were settled. This I did. He wrote an article for the paper about our office, the area we covered, our purpose and responsibilities, the many farmers our work would bring to Yorkton, and the new families who moved into the city. Subsequently he enquired about our work whenever we met. Snippets of our program status often appeared in his weekly column or an editorial he authored.

On the mast head of the *Enterprise's* editorial page was an adage to which I subscribed: "When everyone thinks alike, no one thinks very much." It had a big influence on me and often came to mind throughout my career when challenging the status quo of my contemporaries, believing I was being ornery when my motive was really to find the best solution to problems we encountered in plowing the new ground in which we found ourselves.

In 1954 when my wife and I built our new home, we were guinea pigs on the undertaking, as I was the first Second World War veteran in Canada to take advantage of a new self-help housing program launched by the Federal Government that summer. Wynn authored articles about Jean and I, the building lot we bought from the city for $50, the different roles played by the Veterans Land Act, National Housing Act and Central Mortgage and Housing in our venture, plus the progress of my construction efforts.

* * *

WHILE THE THRUST of our work was mainly with farm groups, individual applicants were always accepted and given electricity if their buildings were within half a mile or less from an existing power line. Many who were just over this half-mile limit were desperate to get service and doggedly pleaded their cases. Some tried to stretch the half-mile with fudged drawings of their farm sites in relation to the nearby line, needing an on-site validation by the local DO. When it was verified they were too far they were declined, encouraged to become part of a group we would deal with later when a project could be organized. A few hundred "singles" were connected each year, the construction work being done by the DOs or the district superintendency three-man operation crews.

I was once offered a bribe, the one and only time, of $500 by a Langenburg farmer to sneak him in and overlook the half-mile limit. It was in the Yorkton basement office. I refused. Admittedly, it was tempting, as all bribes are. This one represented two months pay of my then $3,000 a year salary and our family finances were tight, as we had just sold our 1947 Chevrolet to help raise the $500 down payment needed to construct our new home. My old bicycle was pressed into service to replace our car. The Langenburg farmer waited for service until a project in his area was organized.

* * *

THE MOVE TO Yorkton meant we then only met farm groups from our area, but they still came in large numbers and, as had been the case in Regina, often with their elected representatives. The most difficult politician we dealt with was Asmunder Loptson, the interim Liberal Opposition Leader and MLA for the nearby Saltcoats constituency. Initially he threw his weight around in a high-handed fashion, but his bark was worse than his bite and our experience with more impatient and detestable politicians kept him easily under control. Our relationship soon became amicable in spite of his disdain for our bosses, the CCF government of the day.

The Old Power Plant

THE YORKTON POWER PLANT was built at the turn of the twentieth century to serve the city. Later a line was extended south-east to supply the town of Saltcoats and a few farms along its route. Initially the plant and city distribution system were municipally owned but had been bought and operated by different private interests until purchased on January 1, 1947, by the Saskatchewan Power Commission from Canadian Utilities Limited. By 1953 its generating capability totalled 2,100 kilowatts (kW) produced from several old and outdated, temperamental, diesel-fuelled engines, 4-cylinder monsters, some that were relics. The newest and largest generator was powered by a 16-cylinder Fairbanks Morse engine.

The combined SPC system capacity of all its power stations at the time totalled 161,900 kWs, so the relatively small Yorkton plant amounted to just over one per cent of the total. During the Second World War a 69,000-volt (69 kV), high-voltage transmission line, of single pole configuration, was built north from the Estevan Power Station to supply substations at several places along its route, which terminated at Yorkton. After the line's completion the Yorkton plant operated only in a standby role. By 1953, because of incessant system load growth and new customer acquisition, including the farms we continued to add, this high-voltage line carried more load at peak times than its design capability. When this overload condition prevailed, usually in the winter months, the engines were started, one after another in mid-afternoon and ran through the supper-hour peak load time to help maintain adequate voltage conditions in the city. The colder the weather and heavier the demand by customers the more units were put on-line by the plant's superintendent, Doug Nurse and his two staff.

* * *

POWER FAILURES and inadequate voltage conditions were prevalent in Yorkton in the early 50s, more often than SPC employees, including the local operation managers and staff, liked to admit. The local media was justifiably critical of this questionable level of service, as was Mayor Mickey Novak, City Council and the business community. Their complaints flared up after every blackout. It was no longer adequate for utility staff to simply say, "That's the way it goes," or "We can't control nature," or, "The equipment failed and we don't have spare parts." Electricity had become a more essential service in the daily lives of people. The public expected and deserved a more reliable quality of service.

On July 5, 1954, came the most serious outage that affected Yorkton for many years. A tornado blew down over a half mile of the single pole 69 kV transmission line north of Grenfell and, as luck would have it, the tangled mess was in the middle of a large slough. All the north-eastern part of the province beyond the downed line to Melville, Yorkton and beyond to Churchbridge, Ituna, Canora and Preeceville was plunged into what would be prolonged darkness. By the time a heavy line construction crew was mustered, material moved on-site and the downed line rebuilt, the interruption lasted over 20 hours. Even though the Yorkton power plant engines and generating units were pushed well beyond their rated capacity, it could only handle about a third of the city at any one time, even considering the light summer demands. City areas had rotating blackouts, in the dark about 40 minutes of each hour. The many communities and rural areas normally served from the Melville and Yorkton high-voltage substations were out of service the whole time, although after midnight the town of Canora was picked up for few hours as their badly needed hospital's standby generator had malfunctioned.

Helpful announcers on radio station CJGX encouraged electricity conservation, repeatedly asked listeners and downtown stores to switch off non-essential and unneeded lights, motors and equipment so more customers could be picked up in the hopelessly inadequate rotation scheme. Jessop and I from the Farm Branch and others of the operations staff were recruited to visit every commercial user: cafes, stores, warehouse and shops. Cooperation was good from nearly everyone, except for Safeway's, a big user, whose belligerent manager refused any assistance. He sang a different song when the ice cream in the store's freezers started to get ready to run down the aisles, and no sympathy was given when he stormed into the office and demanded more frequent service rotation than was the store's fair share.

The local staff thought they had given yeomen service to help consumers through this emergency, but the paltry efforts, mine included, fell far short of what was expected and reasonable from their power company. It was embarrassing to be a Corporation employee when confronted by friends and acquaintances when the inconveniences they experienced during the outage were mentioned. Bert Allen, the Yorkton District Electrical Superintendent should have been more assertive with his line of authority to senior levels of the Corporation about the overloaded conditions and frailty of the tenuous old transmission line, and years earlier insisted on an alternate source of power into the area. Apart from Allen's inability to think big, his only defence was his former indoctrination in earlier days with the investor owned Canadian Utilities. It was customary for the private power companies to delay the costs of any system improvements as long as possible to wring out every last cent from existing facilities before the needed changes were made to keep the bottom line as big as humanly possible, even at the expense of adequate service to customers.

In the weeks that followed this outage considerable public criticism was levelled at SPC with intense political pressure unleashed from the local media, (Sam Wynn amongst them), city councils of Yorkton and Melville, the many other towns an villages, their chambers of commerce and the MLAs who represented the affected areas. The condemnation was fully deserved. An alternate source of supply and a more reliable service was justifiably demanded. In only a matter of weeks Head Office announced plans for a new 138,000-volt (138 kV) transmission line, labelled the Red Jacket line. It would be built from Estevan to Yorkton, via the hamlet of Red Jacket near Moosomin, along with a new substation and switching station south of the city. This would give a secure, alternate supply but, more importantly, increased capability to handle the growing loads in the region. By year-end 113 miles of this new line was in service as far as Red Jacket. The new substation there picked up the 25 kV distribution line that served all the communities and farming areas from Grenfell east to Moosomin, along #1 Highway, which helped to relieve the loads on the old 69 kV transmission line so more capacity was available to the Melville and Yorkton substations. That winter the Yorkton Power Station engines ran more frequently and for longer hours to bolster the still overloaded 69 kV line until the balance of the new facility to Yorkton from Red Jacket was completed in 1955.

In 1957 a 138 kV transmission line was extended from Saskatoon to this new substation and switching station. This added greater stability to the high-voltage grid system and gave the north-east part of the province another alternative source of power which, up to then, had relied entirely on the coal-fired power plants at Estevan.

* * *

THE FARM ELECTRIFICATION Branch Office and staff needed a better workplace than the basement hovel we endured. The announcement of the new transmission line meant the old power plant days were numbered, even in a standby mode. In the fall of 1954, after the big power failure, with commercial rental space in Yorkton still hard to come by, the Farm Branch moved into what had been the office and spare parts room on the power plant's second floor. When cleaned and painted it was better than the District Office basement, but still marginal by any standard.

Doug Nurse's surly personality telegraphed his displeasure and objections to our presence and the renovation decision which gave him a small office next to ours. In fairness we were invaders of his space which had lost its relevance in the scheme of things. His feelings were understandable but his resentment was unpleasant. To help appease him I had Barb Hodson do his typing and answer his telephone which helped a bit. Frank Crandell and his engineering staff were then isolated from our front office where the stream of farmers continued. If the plant's generating units happened to be in operation when a farm delegation was there that winter, being mechanically minded, they wanted to see first-hand how electricity was produced in a big way. So we gave mini tours to the plant's generating floor with Nurse's reluctant concurrence.

Dependent on how many generating units needed to be fired up that winter the noise in our offices could be uncomfortable, in addition to the floor vibration, the rattle of the windows and the odours of engine exhaust, diesel fuel and lubricating oil. But we could at least open the windows for fresh air. All this evaporated when the new transmission line came into service a few months later and the plant's engines ran for the last time.

Our poor working conditions were common in many other areas of the company as new customer growth of both the electric and natural gas systems continued to almost explode, year after year, and the Corporation struggled to keep pace with decent office accommodations. By the end of 1954 the number of SPC employees surpassed the 1,000 mark.

That fall my boss, Garnet Parcher, was diagnosed with inoperable intestinal cancer. It was a sad and painful time for me. Working with our desks close together, we shared work details and he participated as best he could, but often blankly stared out the window for lengthy periods trying to come to grips with his mortality. None of us had experience or knowledge on how to support a colleague in this condition, but I think he appreciated our acceptance of his presence as we kept the office running smoothly. I substituted in his position for over a year with the field and road work for which he did not have the strength. On nice days in the spring and summer of 1955 I often took him on short day trips to meet committees of nearby projects. He enjoyed those jaunts as he could get his mind off his physical condition and talk about his life experiences and office issues away from the rest of the staff.

One afternoon I took Garnet just south of the city to watch an SPC heavy line construction crew erect structures for the new 138 kV transmission line. With a crane, the three-pole H-frame structure, two 50- or 55-foot poles connected by a 30-foot spar on which the three strings of insulators were hung, were lowered in the ground simultaneously. As a former lineman, going back to the days before mechanical hole-digging machines came on the scene, he was fascinated with the process and enjoyed the day. That fall my friend and boss died.

My substitution role for Garnet ended and Harry Jessop took over the senior field rep position and the responsibility for running the branch office. The junior field rep job opened for competition. I bid, was successful and, on November 16, 1955, was appointed. Another rung up the ladder out of the union's protection and into a so-called management position, only because there was a lot of overtime involved and managers didn't get paid extra for working long and late hours. My hourly working rate of pay decreased significantly with this promotion.

The Yorkton Farm Electrification Branch Office remained in the old power plant building until the District Superintendency Office moved several years later from its downtown Third Avenue location to a Broadway Avenue shopping mall, not far from the old plant. There the Farm Branch had half decent space after 10 years of very shabby digs.

* * *

AS OUR POWER LINES snaked their way across the province hundreds of towns and villages were electrified. Many of these communities had limited electrical service produced from small generators driven by diesel and gasoline engines. SPC acquired these engines and generators and shipped them to Yorkton where they were stored in a large metal warehouse adjacent to the power plant. This was the provincial repository for this used equipment. When Nurse's power plant was quiet he and his men overhauled those units that were in reasonable condition. SPC's commercial reps sold these refurbished engine and generator sets to meet emergency standby electricity needs in hospitals, confined poultry and livestock barns, hatcheries, underground mines and to others where an uninterrupted supply was essential for the business. But most were dismantled by Nurse and his men for spare parts, or were simply scrapped.

These used units were a hodgepodge of makes and models and sizes that had been kept running by a plethora of dedicated and innovative operators in the communities from which this equipment came. There were many stories of devoted mechanics that had not taken a vacation for 20 years until arrival of our lines and their little plant was shut down for the last time. In the early days of electricity these plants usually only ran from sundown until maybe 10 o'clock in the evening, plus Monday morning for wash day chores and Tuesday afternoon for ironing duties, a fascinating admission about electricity's benefit to the women of the community. One tale emerged about an ingenious operator who gerry-rigged a tin can connected to the diesel fuel line feeding the engine. At 10 o'clock, the plant's normal shutdown time that plunged the town into darkness until the morning, he would turn off the engine's main fuel line and open the one from the tin can that held a cup or so of fuel, just enough to keep the engine running for about 20 minutes. Then he put on his coat, locked the plant doors, walked home, prepared for bed with the lights on, then crawled under the covers just as the can emptied. The engine and generator then ground to a halt and the town went dark. "Cool," as the kids would say today.

The Union

AFTER THE YORKTON move and with Garnet Parcher away much of the time, responsibility for running the office fell on my shoulders. This was significantly different than in the Regina environment where I reported to a senior clerk who handled some duties, but many daily routines were handled by other specialty Head Office staff: incoming and outgoing mail, daily office cleaning, telephone equipment needs, banking and money reconciliations, receipt issuing, ordering of stationery supplies and office equipment. I was now responsible for these functions plus the day-to-day relations with farmers who came to see us, telephoned or wrote. We issued and mailed receipts to each farmer for the money collected, made bank deposits and transferred funds to Regina, a significant chore as we handled nearly a million dollars in the first six months we were in Yorkton. As well as supervising the clerical and stenographic duties, I was the day-to-day link between the engineering staff, the farmers and Head Office. I believed my work deserved more pay.

With Parcher's strong support I rewrote my job description to reflect the new and changed duties. My reclassification request went to the IBEW for discussion and resolution with SPC's management. A short while later, I knew it was under review when Charlie asked me about some duties included in my rewrite. He seemed surprised at some of my answers, that the work in the field was so different than in Head Office, but made no attempt to downplay their importance to the running of Yorkton office.

The outcome unceremoniously came a short while later in a straightforward memo from the Regina union office, a reclassification several pay levels higher plus retroactivity to the date of our move to Yorkton. My back pay exceeded $1,200, a huge amount when compared to my previous year's $3,177 gross income. With overtime there was now little difference between my salary and the rep's. Parcher, a former staunch union member, was delighted. Jessop was openly resentful. Glen Gorham, the Yorkton commercial field rep was furious. I was ecstatic. Had I raised this matter with Charlie it is likely a pay raise would have come, but likely not with retroactivity. And with the union's involvement, the raise was, without any doubt, higher. I had just moved up another rung on my personal success ladder.

* * *

IN 1954 OTHER MEMBERS of the Yorkton local of IBEW felt the presence of the Farm Branch office staff warranted executive representation. I was railroaded into the vice-presidential role which did not thrill me, as there were some union issues incompatible with my principles, in spite of my acute awareness the workplace was far better than it otherwise would have been had unions not pioneered and fought for the needed changes and improvements. And, of course, I could not overlook the benefits my recent reclassification had brought to me and my family. But my union role ended after almost a year of incessant hard work to construct a new home by myself, working every evening, on weekends and during my accumulated holidays.

In 1955 company contract negotiations with IBEW seriously foundered. When a deadlock was reached the union's Canadian director, Neil Reimer, entered the scene and negotiated directly with Premier Douglas, an unprecedented situation, later described by others as perhaps the most serious labour dispute Douglas faced in his years as premier, including his 1962 confrontation with doctors over medicare.

To break the stalemate IBEW obtained a strong strike

mandate from its members. During this period of unrest and to abet the union's cause, two organizers, Alex McAuslane and Cy Palmer, came from other provinces. McAuslane, a wily veteran of many bitter labour disputes who had even lost an eye in a violent Winnipeg strike, came to Yorkton on a provincial tour of local union branches to explain their side of the impasse, to garner member solidarity and conduct a strike vote, he said, "Just in case it was needed." With the uncanny ability to sway opinions, I thought he stretched the truth when it suited him, to foment relations against the Corporation.

During the meeting, as the discussion centered on a strike vote, we were told if negotiations broke down and a walkout occurred power failures would likely occur but that union linemen would be allowed to work during those outages if hospitals were in the dark or case of other emergencies. My nature to avoid conflict and to compromise prompted me to strongly argue against the strike vote. I was not about to jeopardize my hard work at both the office and at home. We had sold our car and made other sacrifices to finance and build our new house where electricity was essential for heat and food preparation. This was not a dispute with my wife and two young children and the customers we served. George Sinclair supported my arguments. McAuslane glared his displeasure at us knowing the unanimous support he sought was now not likely. The secret ballot vote was all in favour except for two. The displayed resentment made me worry about getting home safely that night, but I did.

Negotiations went to the limit and then to arbitration with a threatened province-wide strike hanging overhead. Premier Douglas's appointee to arbitrate the proceedings was David Lewis, an Ontario labour lawyer, later the leader of the federal New Democratic Party, the renamed CCF political party. A new contract was hammered out and the strike averted.

Before the Yorkton local's next meeting they needed a new vice-president.

* * *

OVER THE YEARS as a trade union member and, later, on the management side supervising unionized staff, I repeatedly felt the company was not as thorough as was the union in preparing for contract negotiations. When bargaining broke down and strikes happened or lockouts were imposed I believed it a failure on both sides, but more so on management. If workers have grievances about their workplace, pay scales, working conditions, hours of work or other matters covered in the contracts, or new issues that arise that are not covered, it is management's role to ensure the union and its membership understand why such benefits can or cannot be granted, in whole or in part. Managers have to be open and honest, to make an extraordinary effort to communicate with workers so grievances, unrest, slow-downs, lockouts and strikes can be averted. Many managers were disdainful of unions and were dismayed when their behind-the-back strategies backfired. Then there were industrial relations managers who had been in their positions too long, had grown stale, and detested the union and its representatives, spoke ill of them to all and sundry, then wondered why they were frequently at each other's throats.

In the 1970s Bill Walker, SPC's Industrial Relations Manager and I were friendly and often speculated over coffee about a theory he pondered to stop the game-playing that characterized union-management negotiations. It was for each side to present to an unbiased third party a package of contract items they believed would form the basis for a new agreement. This third party would accept one side's or the other's without alteration and it would become the new contract, each side obliged to live with its contents. It is likely these submissions would be pretty realistic and not the absurd positions about pay scales, working hours, fringe benefits, pension plan changes and so on that continue to characterize the initial ritual dances known as the opening of negotiations. There is likely little chance that any company industrial relations negotiations people would ever consider doing themselves out of a job, or the union international representatives for that matter, who always continued to get paid when the union they represented went on strike. Wishful thinking?

1954 – 1957

Aerial Field Checks

AS WE PREPARED to get the 1953 program up-and-running our mapping needed more accurate farm site locations. The data gleaned from the RM councillors had served its initial purpose to determine the AC prices farmers would pay to get electricity. However, their mapping details of farm sites lacked the precision we needed for two reasons. First, the layout of potential power lines that would ultimately connect all farms was important for engineering reasons and, second, each new project's boundaries had to be carefully drawn. Both had to be defendable as they came under severe scrutiny when exposed to the farm community at a project's organizing meeting. If farms were left out that had no chance of being part of another project, like those on the edge of the boreal forest on the northern perimeter of our area, changes to include them had to be made at the meeting that demeaned the importance we gave them. Also, if the number of habitable farms in a project were inaccurate, adjustments to sign-up targets had to be made on the spot, also not a good strategy to have to do in public.

We had learned from Murray Leach's earlier foray that field-checking by car was impractical. The only other option was to do it by air, hovering over the countryside of each project to gather the needed intelligence. The idea was concocted in the Bunker, and its origin came from a long-used utility practice, dating back as early as 1936, by the operations staff to inspect power lines from the air for damage or, on occasion, to find the location of downed lines after storms.

The inaugural field-check test flight was a huge success performed by the Regina Area's Jim Rutherford and Siggy Bock, in a chartered aircraft from Estevan's Mitchison Flying Service. They were enthusiastic about the information they were able to glean and, as well, had completed half a dozen projects near the American border in a matter of only three hours of flying time. After this triumphant experiment the other two areas followed their lead and checked their next years' projects that summer, when flying conditions were the most favourable and before the fall organizational meetings were scheduled.

Flying about 80 miles an hour at an altitude of 800 feet the flight paths followed the section half-mile lines, going back and forth the length and breadth of each project. From this bird's-eye view of the topography we could see with ease at least a half a mile on either side of the aircraft, out to the road allowances. With clipboards on our knees to hold the maps, scaled one inch to the mile, farm site locations were accurately marked with a system of symbols showing the building conditions as good, fair, poor and not habitable. These judgmental designations were carefully kept to ourselves because there were times when the buildings of our committee members were a bit on the dilapidated side. Rural schools, community halls, inland stores and churches were similarly labelled, the latter shown with an † on its symbol.

The municipal roads were categorized from good through to impassable as it was useful, when possible, to route the rural lines where the DOs would later have ease of access to service them in adverse weather conditions. Patches of bush were outlined as well as the size of the trees, needed to accurately estimate the costs of clearing these trees and undergrowth for the power line right of way, an expense that was part of building that job. Other construction obstacles to line routes were noted, like gravel pits, deep ravines, cemeteries, school or community playgrounds, large sloughs that were wet year-round, lakes and streams. The existence of telephone lines were noted, particularly if they happened to be on the north and east edges of the road allowances, because they normally were built on the south and west sides. If a building location or other details needed verification it was easy to have the pilot turn around to recheck. Several projects could be completed in the time a full tank of gas kept the plane airborne.

First Frank Crandell and Garnet Parcher, and later Harry Jessop and I did the flying in our area with Austin Ingham's Yorkton Flying Service. Their hangars and grass landing field were located at the city's southern limit. Tom Schollie, a University of Saskatchewan law student, was assigned as our pilot and quickly became familiar with our needs. Flying was his summer employment to help with his university education costs. After graduation he practised law in Alberta and by the mid 1990s had become a well respected provincial judge in Red Deer.

Before each sortie Schollie filed a flight plan with the Department of Transport (DOT) in Winnipeg. Amongst other things it listed where they would be flying, their expected time in the air and estimated time of arrival back home. Upon landing, Tom would notify DOT that they had landed and the plan could be closed.

One afternoon DOT called our office and like a bolt-out-of-the-blue asked me, "Do you people have staff flying today with Yorkton Flying Services?" "Yes we do," I replied, "Garnet Parcher and Frank Crandell are the on-board passengers. They are flying over some rural districts we plan to electrify next year" Then he asked, "Can you be more precise about where they were to fly?" "Yes," I said, "They were to field-check several projects south and east of Grenfell towards Moose Mountain Provincial Park." I gave him specific directions for each job in relation to the nearest community. "Well," he continued, "Their flight is overdue. Their fuel supply has long since been exhausted, but as it is such a fine day with so little wind, I am sure they are okay. We will do some more investigating and the minute we have any further news we will call." I was alarmed for the safety of my friends and co-workers. While the man expressed confidence I sensed an obvious tinge of concern in his voice.

No sooner had I hung up my phone when it rang again. It was Austin Ingham from the airfield. He had also been alerted by DOT and suggested, "I don't think we should alert Garnet and Frank's families just yet. I have great confidence in Tom and I think if they were in any trouble we would have heard something by now." Austin and I kept in touch for the next couple of hours but as time passed with no news or a glimpse of the plane our concerns escalated. Finally, with a tone of relief in his voice Austin called, "They've just circled the field and are about to land. Everything looks normal." What great news! It had been such a smooth flying day they had decided to land in a farmer's pasture on the outskirts of Kipling. They had walked into town for lunch, had a service station refill the plane's gasoline tanks, then had taken off and continued to check more jobs than had originally been planned. In the process Schollie had neglected to notify DOT of the extended flight time, hence the panic.

In August 1954, I accompanied Parcher on a day's flight in the Moosomin area near the Manitoba border. Flying conditions were poor, with dark, overcast skies, some rain and a strong wind from the north-west. It was a bumpy ride that I did not relish at all and, besides, it made accurate marking of the data on our maps a challenge. Midway though one of the jobs we flew into a heavy downpour, a cloudburst, so severe the engine coughed and sputtered, then died. For me, sheer panic! Why did I agree to fly that day? Was this the end? What now? Where do we land? There was not much time to think but Schollie's skill and experience came to the fore. As we glided and started to lose altitude he calmly turned the aircraft tail to the wind to reduce the amount of rain entering the engine. At the same time he looked hopelessly on both sides for a place to land, maybe on a narrow municipal road. The water table in the area over which we flew was above ground level, as it had been an unusually wet summer, and the roads below us were inundated. Luckily, the engine's heat dried the moisture sufficiently that on Schollie's first crank of the starter it came to life. We had by this time descended to about 300 feet above the ground. Without any hesitation or approval from us he said, "We're headin' home." He skirted the squall, climbed and headed north-west into the wind for Yorkton. The field-checking on that job was completed on another flight.

Apart from a few airsickness incidents no accidents occurred from this activity. The purpose of the flights became known in the farm communities and often brought forth comments from farmers at organizing meetings about seeing a low-flying airplane over their farm. Inevitably they would ask, "Was that you up there?" They knew the information we had obtained was reliable which removed disagreements over the validity and location of farm sites on our maps and the location of project boundaries, both of which had plagued us in earlier days.

Aerial field-checks continued each year on every project in all three areas of the province until the farm electrification program was completed.

Organizing Meetings

IN SPITE OF the penchant of elected officials to often renege on election promises, the 1952 political announcement of a four-year farm electrification plan caught on and whooshed through the rural community like a raging prairie fire. There was little likelihood the politicos could back away from this one. It raised expectations with thousands of farmers, albeit many were skeptical. We had petitions from more groups than could be handled in the first two years, maybe three, of the program and more kept arriving. In many cases the proximity to power sources dictated their waiting a couple of years or more. Others were disgruntled because groups closer to power sources hadn't yet applied but would get service ahead of them. In spite of these and other reasons of logic the program was well and gratefully accepted by a majority of the farming community. I was one of six farm electrification field reps who relentlessly reiterated the message to eager audiences at hundreds of meetings province-wide, how their farms would get electricity.

* * *

FROM THE OUTSET, to reduce costs, it was planned farmers would do much of the preparatory work, just as they had done in earlier times when they enthusiastically banded together at barn building bees or in other cooperative ventures to erect a school, a community hall, a church or seeded and harvested a crop for an ill or recently bereaved neighbour. Farmers, although intensely independent individually have an abundant supply of horse-sense when it comes time to join forces for their mutual benefit. Next to his family, the closest thing to a farmer's heart is his farm and the land. The advent of electricity would raise his family's living standards and improve the farm's productivity, strong motivation to work together in a cooperative spirit that repeatedly amazed me and became a personal inspiration.

Action plans to get the 1953 program up-and-running started in the summer of 1952. The aerial field checks were completed so each project boundary was carefully drawn and we could tally the number of new farms we could expect to connect that year until we had sufficient jobs to meet our year's target. Organizing meetings were scheduled after harvest in consultation with the local committee chairman. Only under unusual circumstances did we hold meetings in December, a busy time for families with Christmas activities. But they started up again in early January and about two weeks later for those projects where the locals celebrated the Orthodox faith.

Every project had its own meeting. That event launched the farm committee's canvass to coerce and convince their neighbours to fill out an application and pay their money, a time to put-up or shut-up. Some committees performed this work with greater proficiency than others as they more carefully organized their canvass with strategies for maximum benefit, which made our follow-up easier when they had finished. They ensured no one was canvassed by someone who was on the "outs" with the prospect. Other committees barged ahead with little thought, inevitably fell short of their sign-up targets, fully expecting us to salvage their quandary. Others ran into severe, legitimate objections and difficulties that inhibited their success.

When I met the committees a month or so after the organizing meeting to pick up the money and completed forms, the final clean-up stage of their work, and found they were seriously short of their target because of inferior canvassing efforts, it was hard to tell them the project had to be dropped from the program. The money collected would have to be refunded and the group would wait a year or more before it was tried it again. Often it was not just their canvassing efforts which were the cause of their failure but poor crops or low produce prices that year or political animosity against the CCF. Whatever reason, another project slated for the succeeding year was brought forward to replace them.

When these defeats occurred a little bit of me was whittled away in the process. Not liking to accept defeat I lived and suffered with the disappointments of those to whom we had to refund payments. A hastily organized meeting was held with the new group and, if this was towards spring, the sign-up had to be completed quickly. The new folks were excited at the prospect of getting service earlier than they had expected, and usually leaped at the opportunity and seemed particularly appreciative of our efforts. On more than one occasion line construction happened to include the newly picked up group early in their building schedule so the farmers had service only weeks after the organizing meeting. The dropped group was included in our program when local conditions changed or improved and we had reason to believe another try would be successful.

Each year's projects needed to be ready well before spring breakup when the frost came out of the ground and highway road bans were lifted, permitting travel of our heavy construction crew vehicles. In anticipation of their arrival at each job power line material and supplies were delivered and stockpiled in nearby community railway sidings or on a vacant lot or in an unused corner of a farmer's field. At the time of project organizing meetings we had no idea of the construction crew schedules so all jobs had to ready as the crews moved from one job to the next with a minimum of travel distance in between. Construction time was critical, particularly if wet weather delayed their ability to work, as the Canadian prairies only have about 100 frost-free days a year.

* * *

ORGANIZING MEETINGS were usually held in the evening in one-room country schoolhouses that dotted the lonely prairie countrysides. Less frequently we gathered in a community hall where they existed in some districts, or in a church hall or basement, sometimes in the church's sanctuary itself, or in a large living room of a committee member's home. Meetings conducted in a church sanctuary took on a more sombre ambience. Preaching about electricity and its benefits from a pulpit seemed a bit out of place, particularly if tempers flared or the language deteriorated. I once held a meeting for about 20 farmers, all men, on a school bus parked on the roadside while a blizzard raged outside. These country locales within the project's boundary were preferred because the meeting's focus would be only on that job. If held in a nearby community theatre or town hall, attendance included not only the intended group, but farmers from every direction outside the designated project plus curious townsfolk, there for whatever entertainment value the event might offer. The number of people at these larger sessions often exceeded 200 when perhaps up to 50 or 60 were expected. They lasted longer too, well into the night, as many questions came from those annoyed at their exclusion from the project or maybe even from that year's program. Also, it was inevitable with such a ready-made large audience that some aspiring political advocate, maybe a local MLA, either government or opposition, it made no difference, or some other tub-thumper, could not resist the temptation to beat his drum for whatever was their cause. It was awkward to deal firmly, yet respectfully, with these crusaders while still keeping the meeting's agenda on track.

On the other side of this coin were the respected community sages, mentally well-heeled narrators, who moralized on what we were about, urging the audience to come together in a co-operative spirit and get behind the farm committee for a successful sign-up and to get it done quickly. These less frequent orations always pleased me and they were thanked at the time or, privately, at the meeting's end.

A large percentage of attendees smoked, mostly roll-your-owns but a few puffed on tailor-mades. As there were no restrictions on the habit, except in a church sanctuary, the odour and foul air in crowded spaces where ventilation was inadequate was not a problem for me as I too was a smoker in those days.

Without exception the attendees at my meetings were predominately male. Some wives came, perhaps up to a quarter of some audiences, as did a few older children, curious to witness a bit of history that would be made that night affecting their corner of the world. Not once, in my experience, was a woman named the chairperson of a farm electrification project committee and only twice were ladies named to a sign-up committee to canvass other neighbours, an unmistakable sign-of-the-times that the rural community was a male dominated world. Perhaps this supremacy was a carryover from other cultures or the fact that the women's equality movement had yet to emerge. I often wondered whether if women had had a greater status in contemporary farm society, with greater power and control over their lives, would obtaining electrification subscribers have been easier? As shall be seen in what follows, when I entered the picture to try to convince a reluctant prospect to sign-up so the project's required number of new customers could be achieved, it was always the farmer, not his wife, to whom I made my sales-pitch or argued my case. He made the decision, he signed the contract and the cheque. It was highly probable that during these tussles she was offering up a silent prayer that her husband would succumb to my power play. Maybe I only saw cases where the farm women were sidetracked into silence, almost as if the men depended on that silence for their own so-called success. Where women had a stronger voice in the farmyard, likely that farm had already signed and I had no reason to see them.

The stark reality and the abiding truth of the matter was the farm wife would benefit significantly more than her husband when they got electricity. First and foremost, it would likely abolish outdoor plumbing and its associated Eaton's catalogue, and this needs no elaboration. It would relieve her from the back-breaking labours that gives rise to the often heard expression, "A hard-worked farm wife old at 30." Electric ranges would eliminate hearty meal preparations on hot days on the even hotter coal and wood stoves, particularly for large crews of men at seeding, branding, haying, butchering and harvest times. The twice daily routine of cranking the cream separator could be replaced with an electric motor. Food preservation would be easier with refrigerators and deep-freezes. An automatic clothes washer would replace the gasoline motor driven machine that needed the kitchen door wide open to exhaust the engine's carbon monoxide fumes, a miserable chore for her on a cold day wearing an overcoat or parka. An electric water heater would rid the wash day chore of heating snow-melt water on the stove. Wrestling the stiff as a board frozen long-johns off the clothes lines could be overcome with a clothes drier. Electricity would banish to the attic the gas lamps and kerosene lanterns that were so fire prone. Then there were the advantages of better lighting and a host of small appliances and gadgets to ease her daily burdens. And finally, wasn't an easier and better life deserved after she had laboured under such difficult conditions during those times she was pregnant? Unquestionably, the greatest benefactor of farm electrification was "The Missus."

* * *

IN THE EARLY 1950s, country schoolhouses were still the focal points of prairie culture and the districts social centre. They were the community art repository for live theatrical productions and musical concerts, the core of local science and technology, a gathering place for debate of local issues, a sports centre for baseball games and sports days, a polling place where election campaigns concluded with the casting of ballots and fund-raising events, called "socials," when rag-dolls, boxed-lunches or neckties were auctioned, and where dances and parties went well into the night. Country schools were built on the "school quarter," the quarter section of land assigned by those responsible for the surveys in the interest of the community's education. The school quarter was usually rented and farmed by a neighbour, the funds from which helped financed the school's operation. Each had its own colourful history of distinguished students and events and teachers, and each was descriptively named – Avalon, Bay Island, Beaver Valley, Crocus, Friendship Hill, Hilltop, Lakeview,

Meadowlark, Poplar Bluff, Prairie View, Red Robin, Stirling, Sunkist, Sunny Slope, Valley View and Wheatland, are but a few.

It was a full week's work to conduct five evening meetings and during the day visit farm committees of other nearby projects already in the sign-up stage or explore reasons why a district had not applied for service or to deal with other issues related to our work. Sometimes two meetings were held the same day if their locations were close together. If this happened these were long, gruelling days for me. Some days I would have liked to have had an extra couple of hours, but was stuck with the usual 24.

After the farm project committee chairman decided the meeting's date, time and locale, our office prepared and sent colourful posters for display on notice boards in post offices, curling and skating rinks, elevator offices, stores and cafes. On the meeting day a visit to the local rural telephone office was useful. The operator's general rings on the party lines that cost us a dollar, reminded listeners of that evening's event. It was also useful to visit with at least one of the committee members, to learn of any local or contentious issues that should be avoided at the meeting. If the committee was made up of perhaps not the most respected community people, more enquiries had to be made on the best way to deal with this matter. We would then see the reeve or secretary of the RM, mayor or clerk of a nearby town, the local MLA, an elevator agent, a priest or a minister. It would have been foolish not to meet beforehand with the chairman of a nearby previously electrified project or two, like George Lincoln in the Wawota area. These community leaders had valuable insights into the dynamics of the people in the area we planned to electrify.

* * *

I TRIED TO ARRIVE at the schoolhouses well before the advertised start time, when the classroom was dead quiet except for the ticking of the wind-up pendulum clock above the teacher's desk on the front wall. Almost always on cool or cold nights someone had preceded my arrival to turn up the gravity-fed oil space heater or put on a coal fire in the stove, some that were of the pot-bellied variety.

The evening's stillness was shattered when I started the 1,000-watt lighting plant in the back of the panel van, an overt demonstration of the service we promoted. An extension cord was run from the unit through a window. A couple of 200 watt light bulbs, each encased in metal trouble-light meshes, were plugged in and suspended from overhead hooks normally used to hang the school's gas lamps. I posted project maps on the blackboard, side wall and at the back, being careful to not disturb student art or other displays of their work. The Bell and Howell 16 mm movie projector, its stand and a viewing screen were lugged in and set up. Those who came early were engaged in friendly conversation. I encouraged folks to acquaint themselves with the project boundaries on the posted maps so they knew beforehand the area that was included, who was in and who was out. Shortly before the scheduled start time I cranked up the projector and showed electricity oriented movies like *Running Water on the Farm* or *Making Green Hay* to whet appetites and allow latecomers to get seated.

Often there were amusing moments when burly farmers, or their ample wives, wedged, tucked or slithered themselves into the smaller children's desks in the front rows which were occupied by the most junior grade students in the daytime. Some didn't try or couldn't make the manoeuvre, leaving them little choice but to sit on wooden benches, if the school had them, or stand at the back or along the room's side walls, shifting from one foot to the other throughout the long evening.

If the panel van lighting plant was not working or being repaired or I had a rented vehicle without this equipment, Aladdin gas lamps provided the lighting for those meetings. Their incessant hissing was a distraction for me as I had not grown up with or was used to them, and they did not always function trouble-free. Moths and insects hurtled suicidally into their mantles, which then needed replacement with time lost and friendly jibes from bystanders to those making the delicate repairs with their large, callused fingers. Sometimes the lamps ran out of fuel and when the school's supply can was empty a close neighbour hurried home for more so we could continue.

SPC's operating staff rarely attended these meetings, but I had two exceptions who always came when I was in their districts. These were Verne McQuarrie and Vince Frederick, the DOs from Wadena and Kelvington, respectively. I appreciated their support. Not once did a district electrical superintendent attend one of my meetings even though they always knew we were in their territory. It was understandable that with many meetings in their superintendency each year they could not hope to attend even a few of them but in many instances the project's success meant a new 25 kV line would complete an alternate source of supply, a significant benefit in their work to provide more reliable service to the hundreds of existing customers for whom they were responsible.

It was not unusual for some SPC's operations staff to quietly, and at times not so quietly, oppose the expansion plans of the electric system, as it meant more work for them: more line to maintain, more records to keep, more customers to administer, more meters to read, more unpaid bills to collect. Our work went ahead in spite of this underground resistance. It was understandable I suppose, but toiling behind their backs was awkward and unpleasant. Corporation management should have been more assertive to get these people on-side, but such is the price and the way of progress.

* * *

THE CHAIRMAN of the farm committee, initiators of the original petition wanting electricity usually presided over the organizing meeting. Rarely at these gatherings things deteriorated out of control, but it was proper that someone local be in the chair to keep some measure of decorum. I wanted control of the session to be seen as their meeting, that I was there at the request of the committee. Beforehand, I acquainted him with my agenda and how he could help when questions arose.

After introduction I pursued a casual and informal manner, interlaced throughout with a few jokes. At some meetings humour was not even considered because the entire proceedings needed translation into the native tongue of my audience and time was of the essence. Similarly, I followed the straight and narrow when there was evidence of political animosity against me, SPC or the government or, if some other local issue had tension written all over it. Sometimes I tried to break the ice by starting my remarks with, "Dearly beloved, we are joyfully gathered here today . . ." This usually relaxed my audience, put them at ease for the more serious stuff that followed. I invited questions at any time, did not mind being interrupted, believing it important to satisfy any concerns or uncertainties as they arose. The chairman fielded these questions and made sure they were all dealt with in the order they were raised. This routine and deportment seemed to work well for me.

* * *

THE AGENDA for my hundreds of organizing meetings, and I suppose those of my counterparts, covered all the bases of the farm electrification program. We needed to ensure that during or after sign-up there would be no surprises from any farmer, or that any claim could be made that some aspect had not been carefully and thoroughly explained.

My list of agenda topics were abbreviated to a word or two on one page in a stenographers notebook I kept on a desk or table in front of me. Each night as I waded into them, knowing that it would be a couple or three hours before the last one was retired. I used the school's blackboards to write or draw about what I was trying to describe: number of farms that needed to be signed up; the AC cost less the $50 reduction; the down payment needed from those taking out a loan; the farmyard pole location to be considered a "maypole" from which service wires would fan out to the home, barns, shop, wells and outbuildings; the farm transformer size was usually 3 kVA, or five horsepower, adequate for the average farm but larger ones were available for those whose demands would likely be higher than average; the monthly farm energy rate; the self reading, postage paid, mail-in meter cards; and the appointment of a sign-up committee. All of this was pretty heavy stuff for farm folks to take at face value from a stranger. But thousands of them did, I suppose in part, because of the sincerity with which we preached our message.

* * *

EVERY RURAL COMMUNITY had its own class system and leaders, each distinct from the next. This unofficial hierarchy reflected the ethnic background or the dominant religion of the area, or how long they and their ancestors were in the community, or the type of farming in the district, or the size of their farm acreage, or more visible affluence like the biggest barn or tractor with the most horsepower, or if a person was the local Saskatchewan Wheat Pool delegate, or a political activist, or if they were an Elite seed grower. With little differences from one farm to the next as to the basic labours of farming, there were those who prospered from their hard work. Sometimes just plain old luck separated them from the rest. These leaders were well respected folks and usually spearheaded our committees.

At the end of each meeting the committee members were the last to leave, usually in an optimistic mood, believing the canvass of their neighbours would be relatively straightforward, that the numbers who had signed the petitions asking for service months earlier meant achieving their target numbers was but a formality. Such did not always happen, as we shall see later.

After the last conversation had petered out I unpinned the maps, packed my briefcase, allowed the lighting plant engine to cool down, then, with the aid of a flashlight, secured it in the van, stowed the extension cords, lights, round film can reels, projector, stand and screen. It was usually well past midnight when I headed for my nearby hotel or for home. On arrival, the adrenalin rush of the evening was subsiding. I wrote a brief meeting report, its locale, the bonded committee names and my sense about the job's success. Sleep then came easily.

My audiences were, almost without exception, so eager for my message I felt I was on a crusade. Looking back at it now, I see it was one. Their rapt attention hour after hour, night after night, week after week, to what I had to say about electrifying their farms was exhilarating, knowing my work would forever change their lives. Usually doing the same tasks over and over quickly bored me but in spite of their repetitive nature these meetings and their follow-up was a continuing challenge, and gave me a sense that I had to do this work and get it done.

* * *

IN THE EARLY 1950s the winds of change threatened the very life of the small, independent country school districts, so important in the daily lives of the rural people. It was the emergence of larger school units, joint urban/rural institutions, blessed by the provincial Department Of Education. There were many long, spirited debates about whether or not a district would join the unit. It was a devastating blow to the independence and pride the rural community had in their local school. Part of this new controversial scheme was the need for children to leave the security of a much loved, one-room country school to the ordeal of travelling on cheese coloured school buses to a larger institution full of strangers, even though these bigger schools had better educational and teaching amenities, convincing arguments in their favour. Some rural schools were kept open for a time but their days were numbered. In many of my meetings the issue of larger school units was a subject to be avoided at all costs.

I had many meetings in country schools that were part of a larger unit. The minute I walked into these classrooms, the benefits of the amalgamation were obvious. The children now had better teaching aids, more modern textbooks, better visual materials and equipment, better lighting and, often, better maintained buildings. There seemed to be a good feeling that permeated through me in those classrooms.

Rural schools in farm projects were almost always electrified by the local school district boards at the same time as the farms. However, as the larger school unit scheme expanded the planned closure of a rural school was often telegraphed to the local people when they learned it would not get the lights, news that was often not gracefully accepted.

After 1956 my meetings in rural schools without electrical service became rare.

* * *

AS MENTIONED EARLIER, my work in the Yorkton Area was with a high percentage of small acreage farmers, immigrants or first generation families that originally came from a host of countries in Europe and western Asia, some being the survivors of oppression and persecution. They had homesteaded and worked the land to create Canada's breadbasket. Many were elderly, some poorly educated because of language difficulties. Many spoke only their native tongue, so patience and empathy was needed in translating our meetings and conversations. It was essential to build a level of trust so I showed them respect, always tried to present myself as an honest and straightforward person and never failed to do what I had promised. Some had at one time or another been cheated out of the fruits of their labour by grain company agents, banks and money lenders, livestock buyers or unscrupulous merchants. Understandably, they were suspicious, unsure about government and its agencies, or commercial businesses.

Usually these were people of few words who said what they meant in a plain and succinct way. One part-time farmer, of Ukrainian descent, on the "Fenwood North" project committee, when I asked his occupation, said "Verk saction Fanwood." Translated, he meant he farmed his land part-time and worked on the railroad section gang headquartered in Fenwood maintaining the Canadian National Railway mainline tracks, switches and railway right-of-way west of Melville.

* * *

I HELD SCORE upon score of organizational meetings between 1952 and 1957. Most were ordinary, unspectacular events, and followed predictable patterns, one undistinguishable from the next, in spite of their almost guaranteed repetitiveness. However, a few were unforgettable affairs.

Some Memorable Meetings

AMIDST THE HUSTLE and bustle in the early fall of 1952 Gord Seel inadvertently booked two meetings for the same evening, many miles apart, for Jim Rutherford, the Regina Area field rep. Rather than reschedule one of them Seel and I eagerly suggested we could conduct the one for "Central Butte South." Charlie and Jim bought our offer. Both had confidence in us and we were champing at the bit for the experience. This was well outside our clerical job descriptions, an infraction in which our IBEW union had more than a casual interest. Our shop steward was, of course, not consulted.

To prepare ourselves for this venture we accompanied Rutherford to a meeting near Regina with the "Lajord South" group a week before our scheduled performance. As the three of us headed east of Regina that night it poured rain, as it had done for a day or two. At Lajord we met one of the Bechard brothers, as had been arranged, in the Co-op Store parking lot. He was from a well known, innovative and progressive farm family from the area, whose farm was in the project to be dealt with that night. The municipal roads were impassable for Rutherford's rear wheel drive car so it was left in town. The three of us rode in Bechard's four-wheel-drive Jeep, capable of maneuvering through the "gumbo" of the Regina plains. Rutherford and Seel, both hefty guys, crammed into the front single passenger seat. I was in the back on an unattached, ten inch square block of wood, with arms outstretched, hands gripping the outer edges of the two front seats with Rutherford's briefcase wedged between my knees. The open-ended canvass cover let rainwater come in unabated, which soon and thoroughly soaked my shivering backside.

About a half a mile west of town we turned a corner and, abruptly through the rain, the headlights displayed a washed-out culvert. Two heavy timbers, about 10 feet long, spanned a rain-swollen creek. Bechard approached slowly in low gear. The front wheels climbed up on the beams, but the rear ones slid off because of the slippery gumbo. Same thing on the second try. He backed off the makeshift crossing and announced, "The only way to get up and across is to repeat what we just did, only we have to do it faster, so the wheels won't skid off the planks." It was quite a thrill as we bounced up and over their ends and to relative safety on the other side. The manoeuvre worked. We had about six miles to go to a rural church, the meeting's locale.

I dried out at the meeting where about 40 people, mostly all men, awaited our arrival. They had walked, or come on horseback, tractors or four-wheel-drives. Rutherford spoke from the pulpit. Seel and I sat in the front pew and made notes for our adventure the following week. The rain continued and the meeting was uneventful. Our return journey was the same except the water in the road ruts was deeper and the level of the creek higher, but the earlier bold tactic on the planks was again excitingly successful. We arrived safely. I was soaked again.

The following week Seel and I checked in to the Central Butte hotel, had supper, then went south to the Red Robin, a one-room rural school house, the meeting's locale. It had been a warm, sunny, windless, fall day. The classroom was like an oven, the air inside was still and stale. In preparation for winter the storm windows had been screwed on and were impossible to remove. The only fresh and cooler air in the room came from the open rear door and the three tiny, round vent holes at the bottom of each of the six storm windows.

By meeting time the tiny classroom overflowed with stand-ins at the back and along both sides. On one wall an Opposition MLA led a group of discontented farmers from outside the project. They were in a dour mood as they were not part of the 1953 program. Adding to the sweltering heat were two

Aladdin gas lamps that hissed their stark rays, attracted the usual hordes of bugs and insects hell-bent on their way to self destruction. Seel handled most of the agenda with relief from me to deal with questions as they arose and, after the meeting, completed the bonding forms and showed the sign-up committee how to complete the forms and gave them a pep talk.

Throughout the long evening in the small, front row desk near the teacher's sat a mentally challenged lad who became our greatly appreciated, self-appointed, diligent water-boy. Periodically he would nod at us as if to say, "Need water?" and, with our concurring nod, he took the only small glass outside to the well where the pump handle screeched its objections, piercing the stillness of the night. Then he returned, only to repeat the routine over and over. The meeting was long because of our inexperience and the impatient overtures of the MLA and his grumbling group. The two of us earned our overtime pay that night on this virgin experience, certainly a baptism under fire. We must have done reasonably well as the project sign-up succeeded. The lines were built the next summer. The Red Robin retired its two Aladdins.

* * *

MOST OF THE AUDIENCE who came to my warm, fall meeting in 1954 north of Yorkton were uniligual Ukrainian, so the entire proceedings had to be translated by a committee member, a process that made me uncomfortable as I was never sure the translation was exact or gave the intended emphasis. This process made for a long evening. Because of the warm day the one-room school was hot, stuffy and lighted by one gas lamp with all of its disadvantages. It hung above the teacher's desk, my podium. At a small grade one student's front desk abutting the teacher's, sat a tiny, elderly lady, her head enfolded in a colourful babushka that framed her wrinkled face. She was one of the first to arrive and had gone directly to that seat. She had not said one word all evening, just sat quietly, listened and watched, her hands folded on the desk top.

When the meeting's post-events started to wind-down and as I started to gather my paraphernalia after most people had left, she rose from her seat, cautiously approached me and intently looked up into my eyes with a sense of hope and determination on her wizened countenance. Taking my hands in hers that were strong and firm, tanned, calloused and gnarled from countless hours of hard work, she intently said in heavily accented English, "Meester, Meester, vee gotta gat dee lites." I asked, "Why is it so important to you?" Without hesitation she replied, "For dee keeds. For dem to study. No more gas lites. No more fires. No, not for me. For dee keeds." She lived with her son, one of my farm committee members and his family. It was her grandchildren about whom she spoke. Assured that I would do all I could for the project's success, she thanked me, patted my hands, smiled warmly, then quietly turned and left. She had patiently waited for over three hours and in those few seconds had determinedly made her case, her contribution to initiate change and improvement for her family and the district.

This brief encounter made a huge impact on me. Her passion and the look on her face resurfaced in my mind in glowing technicolour countless times throughout my life. "Vee gotta gat dee lites," *To Get The Lights,* became an intense force in my life and daily work, hence the title of this memoir.

* * *

FOR WELL OVER a year through the summer and fall of 1954, until late fall of 1955, I substituted for Garnet Parcher when he was too ill to travel. While I often used his panel van with the lighting plant in its back end for meetings, I sometimes had to use a rental vehicle, which in Yorkton was then only available from a taxi company. The car assigned to me had the usual "Taxi" lighted roof sign and the side doors and trunk lid carried its company name and telephone number.

I had this unit for two meetings on successive nights. The first, for the "Fenwood North" job. The second a few miles away, near Jedberg. They were routine, uneventful meetings but nearby local events were not. Earlier that week Mike Petlock of Fenwood murdered his mother, brother and his wife and their two children, allegedly because of their unending abuse towards him. Having eluded police, he was still on the loose the night of my two meetings. Attendance in the country schoolhouses was predictably lower than expected as the locals were as uneasy as me to be out after dark under these conditions. Petlock was apprehended a few days later in Edmonton. He was defended by a Saskatoon lawyer, Emmett Hall, who gained more public prominence later for his study of health care for Canadians. After Petlock's trial and conviction Hall's legal fees gave him title to the Petlock family farm.

The second night I went to the meeting locale in the rented taxi with its boldly illuminated roof sign. I arrived a bit early so before going into the schoolhouse I parked close to the building's east side so the car would be cooler in the shade from the setting sun. In the school's teacherage, about a hundred yards away, I noticed a young couple intently peering at me through parted window curtains. I did not pay them much attention. When the first folks arrived for the meeting I got out of my vehicle to talk with them. As soon as I did the young teacher ran towards us. Obviously irritated, he yelled and swore at me, his wife in pursuit doing the same. He had forgotten there was a meeting in the school that night. When they saw the taxi "hide" beside the school, their imaginations led them to think I was the Petlock family murderer awaiting darkness in a stolen car. He said, "My gun was loaded and ready for you if you had approached the teacherage." They had felt very vulnerable until they saw familiar people talking with me in a non-threatening way. As they vented their understandable anger my shamefaced apologies seemed a bit limp-wristed.

* * *

MY COLDEST NIGHTTIME organizing meeting was with the "Choiceland North" group in that town's community hall. Their chairman, a Mr. Hildebrand, was about my age and we half-knowingly eyed each other all evening with the sort of look that says, "Do I know you? I think I do from somewhere." At the meeting's end when we were able to more freely talk we discovered we were grade one classmates 25 years earlier in Assiniboia. His parents had farmed close to town from where he and his siblings had walked to school. They had been dried out and driven from the Dust Bowl in 1932 during the Great Depression. With tinges of anger at nature's treatment of them and their land they had abandoned their farm and trekked north to live up in the bush, over 350 miles, filled with despair and uncertainty, their worldly goods and family members piled on a horse-drawn wagon, their cow tethered behind. After settling on the frontier of the boreal forest they cleared and broke the land, worked hard and their farm prospered. Because the outdoor temperature was about 50 below zero that night and everyone was anxious to get home we put off our boyhood reminiscences until I came back a month later to help with their sign-up work.

Our vehicles in the parking lot were left running throughout the meeting but even with the warm engine the wheels on my car resisted any movement. Finally, after rocking back and forth it broke free. As it nudged forward the frozen tires were like square blocks until rounding themselves after driving the few blocks towards the highway. At that midnight hour and because of the severity of the temperature, albeit with no wind, there was no other traffic as I reached #55 Highway and turned east towards Nipawin. Not the safest place to be in those conditions, my insides were in a tight knot of anxiety. After only a few miles the engine sputtered to a stop, its gasoline supply plugged with a frozen line somewhere between the tank and carburetor. This was a common occurrence I had experienced many times. I coasted to a stop at the side of the highway. My trunk's emergency kit contained cans of gas line antifreeze, methyl hydrate, an almost sure-fire remedy. Part of a can-full went into the gas tank, the rest for the carburetor, poured into the loosened screw that secured the top of the air cleaner. After a few cranks the motor coughed, started, and when running smoothly I headed the last few miles to the safety of the Avenue Hotel and my warm bed.

* * *

MY MEETING TO organize the "Brewer East" project was in a cavernous rural community hall on another cold and stormy night. This farm committee was a lackadaisical bunch. I arrived at the hall ahead of everyone to find it as cold inside as it was outside. None of them had bothered to go there ahead of time to start the furnace. The first to arrive did light a fire in the pipe-less furnace which took a while to exude warmth from its four foot square, iron floor grate in the center of the auditorium. By the time the meeting was over the hall's temperature was barely above freezing. I staked out the warmest spot in the place, on top of the grate, and throughout the meeting wore my toque, parka and boots.

That meeting's crowning touch was the long night I spent in my car, chilly and snowbound. Before leaving the hall I voiced concerns about the road condition, the wind that was blowing stronger as the evening wore on and with it the drifting snow, and my ability, or inability, to get home safely. It was, I thought, agreed I would call the committee chairman from a pay phone at the Hilltop Curling Rink on #9 Highway, from where it was quite certain I could then make it safely home to Yorkton. If I did not call they would know I was in trouble and were to come to my rescue. I left and four or five miles from the hall became snowbound. I dug the car free and cleared about a 30 foot path, used it as a runway to get a little speed to get through more drifts up ahead, only to bog down again. After several of these manoeuvres for maybe a quarter of a mile it seemed hopeless considering the distance still to go to the highway. I waited in sublime hope, expecting to be rescued. I did not get to the curling rink. I did not make the telephone call. I was not rescued.

I could not see any farm buildings or any light that might have meant shelter or help. I did not dare leave the car in the wind and reduced visibility to seek assistance without knowing where it could be found since it is dangerous in cold weather to risk losing body heat. With it diminishing, so does functioning of the mind, and with its nimbleness slowed down bad decisions are likely. So, I survived the night in my emergency survival gear, ate a couple of chocolate bars, ran the engine intermittently to conserve gas and warm the inside, dozed a bit and waited it out. As dawn started to undress the night I saw a farm house about half a mile away with smoke coming from the chimney. I trudged there through the field for help and, after a warm cup of coffee, the farmer's tractor dragged me to the highway. I paid him and went home. This was the first and only time any farm group or farmer ever left me in the lurch, a most uncharacteristic experience.

* * *

BLIZZARD CONDITIONS prevailed the night of a scheduled meeting with the "Pleasantdale West" project. Drifting snow would certainly block the rural roads. If I was able to even get to their school I would not likely get out that night. At dinnertime I telephoned the chairman who confirmed my suspicions about the road, but the meeting was important for them and they did not want it deferred. They had ingeniously decided to come to me rather than me to them. One neighbour would clear the roads with his tractor-mounted snowplow for another to follow with a school bus to pick up others and we would have our meeting in the bus at the junction of the highway and their rural road. I easily made it to this intersection on #6 Highway and we did our thing inside the bus while the blizzard howled its fury outside. I stood at the front, faced about 20 men jammed together in their seats and proceeded through the agenda quickly as the driver kept the engine running to keep us warm.

An RCMP highway patrol car stopped and the officer wondered what a school bus load of men were up to on the roadside on such a wild night. His query of, "What's going on here?" was responded to with the driver's reply, "We're having a farm electrification meeting of course!" The constable left shaking his head in disbelief.

When we had finished I returned the 30 miles to my room at Melfort's Ozark Hotel, the plow and bus delivered my audience to their homes.

Another storm hit this same area when the committee finished their sign-up. I was to meet them after lunch that Friday afternoon. On my way there my car radio announced a province-wide blizzard alert. Heavy snowfall, high winds and a cold front were moving eastward. North Battleford was already in its grip and it was expected to arrive in Saskatoon about mid-afternoon. I drove to the chairman's farm, met the committee, checked and gathered the forms and money, had a quick lunch and hurriedly left, hoping to outrun the storm and get home for the weekend.

It was dead calm as I headed for Watson in the dark, about four o'clock, but heavy snow had started falling. I had no sooner turned east on #5 Highway towards Wadena, where I had planned to gas up and have an early dinner at a favorite eatery, Rusnell's Cafe, and be home before my children's bedtime, when the storm front arrived with a vengeance. It was like being engulfed all of a sudden under a Hudson Bay blanket. I continued through the blizzard until midway between the two highway turnoffs for the hamlets of Clair and Paswegan. Driving winds were whipping the fallen and falling snow into a frenzy. Visibility was only a few feet. Highway snowplows of the day, not of the rotary type, simply pushed the snow into high banks to each side of the road which left a narrow one-way trail between the walls of the V-shaped cuts. These were already starting to fill. In the poor visibility I poked my car's nose as careful as I could into one long cut, perhaps a couple of hundred yards, only to have it ride up on top of the drifted snow in a way that from my experience I knew I was stranded, stuck solid.

I stripped down and dragged on my long underwear, wool sweater and socks, boots, leather mitts, balaclava and parka. It was impossible to open my station wagon's driver-side door as the car was so deeply embedded so I scrambled over the seats and crawled out the back door. There was no point to even get my shovel out and start digging.

I remembered passing two farms about three miles west of where I was stuck, at the Clair turnoff. I did not dare try to find help off the highway unless I came across a lighted farmhouse. The walk in the cold and against the north-west gale in my emergency gear wasn't all that bad, although my watering eyes froze shut a few times and needed thawing out with a bare, warm hand. On arrival at the first farm, they wanted to help but their tractor wouldn't start. Likewise with both truck and tractor at the other farm. However, they did call a Mr. Hoffman who lived about a mile northeast of where I was stuck. He had a small caterpillar tractor with a front-end blade. He agreed to come and dig me out so I decided to return to my car.

On arrival back at my car to await Hoffman I faced a set of headlights coming the other way which shone eerily through the swirling snow about a hundred yards away. I trudged to the car, also stuck, to discover frantic parents from Kuroki and their screaming baby. The infant had swallowed an open safety pin and, on the advice of the tiny Wadena hospital, were headed for Saskatoon where more expert surgical help was available. Hoffman arrived on his caterpillar. When told of their predicament he agreed to escort them back to Wadena. I had planned to accompany them and abandon my car for the night, but as we struggled to get their car extracted and turned around, a bus became stranded behind my car. I went to the bus. The driver said he had ample fuel to last through the night, to keep everyone warm.

Hoffman's caterpillar and the Kuroki family left for Wadena, about 10 miles away. He said he'd come back or send help if he was unable to return. The storm continued with a vengeance, my car became completely engulfed. The bus, a charter, with about 25 young women from Saskatoon, was headed for Canora. In the bus's warmth to pass the time we talked, sang, drank and ate whatever could be found amongst us. By Saturday morning the storm started to subside and help came, the result of Hoffman's instructions, in the form of Highway Department snowplows. Then began the arduous task of digging out the vehicles, of which there were more that had become stranded throughout the night, their occupants having joined us on the bus. My car was located only because the radio antenna stuck up through the snow. It was dark before I arrived home 24 hours late. Presumably the Kuroki baby was okay, as I heard no later reports to the contrary.

Hoffman was a hero!

* * *

BY HALLOWE'EN in 1956, that winter's snow had yet to arrive, but it was chilly and children were trick-or-treating on the streets of Whitewood when I checked in at the town's only hotel in preparation for a meeting that night in a community hall about 10 miles south.

The "Whitewood South" project chairman was Oliver Hogarth, a small, brusque man with a deep, gravelly voice, who bluntly spoke his mind. It was an understatement to say he was not a CCF supporter, a position he made quite sure I was well aware when the meeting was arranged and he was asked to be its chairman. I told him we were not interested in his politics and that we were available to any and all farmers who wanted electricity. While skeptical, he agreed to assist.

Hogarth and his wife liked to dance and there was a masquerade in town that evening. She dropped him at the hotel and we went to our meeting after which I was to deliver him back to the shindig.

About 50 folks attended my meeting. As was my custom I invited questions at any time as we started into the agenda. From the outset one gent near the back, fortified with a few drinks, was determined to derail the event. He interjected on nearly every statement I made, his remarks laced with anti-government catcalls and innuendo. I repeatedly dodged his provocations and repeatedly explained my purpose. Hogarth was on the spot as his political bent was compatible with that of the interloper. While Hogarth relished the political aspersions he knew the disruptions were not helpful. Whetted by his success and vocal support from others nearby the belligerence of this shit-disturber took on a personal flavour. His tomfoolery was offensive. I said so. I warned the meeting if the abuse against me did not stop I would leave and they would wait at least another year, maybe more, for electricity. The seriousness of my obvious irritation only added fuel to the onslaught. As the uproar escalated I said nothing but started to pack my stuff, unpinned the project maps from the walls, folded them, placed them in my briefcase, bundled up the sign-up kits, left the stage and, as if with blinkers on looking straight ahead, determinedly walked down the centre aisle towards the hall's back door.

The audience fell silent, like a bunch of yokels awed at a circus side show, only then realizing that I'd meant what I'd said. Near the door I sensed some unease amongst the crowd as they well knew the likelihood was very real of their continuing to live with Aladdin gas lamps and kerosene lanterns. At the very last minute Hogarth's voice boomed out, "Anderson, get back up here. We'll stop this nonsense." He asked two men who sat next to their tipsy cohort to remove him. This they did, picked him up by his elbows, despite his vulgar protestations, still audible as he was ushered outside to his vehicle. I returned to the stage, composed myself but did not replace the maps. The crowd knew I was angry. The meeting continued on a sombre and serious vein with nary another interruption. There were no jokes told that night. When it concluded I delivered Hogarth to the party, his wife displeased with the lateness of our arrival as he had missed most of the dance. I went to the hotel, to bed and to sleep.

This was a good farming area and despite the political overtones, easily achieved their sign-up targets and were served the next summer. Hogarth was a good chairman, surprised, yet grateful, politics was never again mentioned in any of our dealings.

In 1971, 15 years later, when I was SPC's Business Manager my wife and I attended a social function in North Portal to recognize our purchase of the village's electrical distribution system from Montana Dakota Utilities, Afterwards, we travelled north on #9 Highway towards Yorkton where I had other business the next day. As we neared Whitewood late that evening the Hogarth farm came into view. The house lights burned brightly. Without hesitation I turned into the yard, much to Jean's chagrin over my audacity to even think about an unannounced visit with a relative stranger at that time of night. The dogs barked. The yard light came on. We were invited in. After a friendly chat and a cup of coffee we continued our journey.

In January, 1992, 21 years after last seeing Oliver Hogarth, and six years after my retirement to Victoria, a lady who sang with my second wife Betty and I in the University of Victoria Choir told us she had grown up in the Whitewood and Round Lake district. Oliver Hogarth was her cousin. She sadly said he had just been diagnosed with terminal cancer. I wrote him immediately of my concern about his health and reminisced about the troublesome Hallowe'en night organizing meeting 36 years earlier. After my letter arrived he called me. It had been shared that morning with his coffee cronies in their hangout cafe. They had laughed over its contents which, of course, brought forth many other yarns.

Four months later, in May, Betty and I travelled from Victoria to Winnipeg in our car along the Trans Canada Highway to visit my stepson and his family. On arrival at the junction of #9 Highway on the eastern outskirts of Whitewood we turned south and again dropped in to the Hogarth farm, unannounced. His wife was in town but he was home, alone, and pleased to see us. Our brief visit was warm and friendly as he shared deep concerns about his mortality, the future of prairie agriculture and his three-generation family farm with no one to succeed him. His only child, a daughter, had no interest in a life in the country. This breakdown within his family of an expected fidelity with the land really troubled him. Sadly, Oliver died seven months later.

* * *

MALE DRESS CODE for office staff in the 1950s was a suit or jacket and slacks, shirt and necktie, necessary in Accounts Payable but more important in the Sales Division as Roy Sarsfield was a stickler on the matter. Packing along a week's supply of white shirts and wearing them in our dusty vehicles while travelling on dirt or muddy roads didn't make much sense to me. As well, our clients never ever wore shirts and ties when I met with them in their kitchens or barns or granaries or workshops or on their tractors; in beer parlours or at hockey games or in curling rink waiting rooms; in hotel lobbies or elevator offices; or in cafes or gassing with cohorts around a Co-op store stove cuddling a can of Players roll-your-owns. I also felt it pretty dumb to be wearing a suit coat while crawling through the strands of a barbwire fence to see a farmer in his field or when I donned my rubber boots to enter a yet to be cleaned out barn of its manure so I could conduct my business or to change a flat tire on a muddy road or to get into the ditch to hook up a tow chain on my car to a farmer's tractor after I had slid off the road. So my wife, Jean, helped me buy two pairs of shirts and pants, wash-and-wear, khaki outfits, worn tie-less. When dirtied because of muddy roads or from changing a flat tire they could be washed in my hotel room at night and be quite presentable the next day. Jean appreciated her lightened weekend washing and ironing chores. And, quite frankly, I thought my outfit looked quite natty.

By 1955 I had perfected the self-help scheme of farmers building the power lines to serve themselves, details of which will be told shortly. Dan Dojack and Ken Barnes, the Saskatoon Area Farm field reps wanted to know more about this venture so they could perhaps apply it in their work. Unbeknownst to me, they, along with Sarsfield, came to an evening meeting I had in Hazel Dell's town hall. The project was a joint farm-town effort, as power to the village depended on the farm sign-up along the way. The farm group committee members were mostly mink breeders, an occupation which fit into our definition of a farm. They would be heavy users of electricity in the preparation and freezing of food supplies for their penned animals.

Many curious town folk attended the meeting. It had been a warm day, the hall was hot and full with more than 200 people. After we had started I noticed three men enter and stand at the back, each in a suit, shirt and tie, sticking out like sore thumbs from all the other folks. It was some time, because of the relatively poor light from the gas lamps, before I recognized them. Immediately to myself I said, "Oh, oh! Sarsfield will sure be unhappy with my khakis."

At the meeting's end, a long one because of townsfolk questions, the four of us talked about self-help details before they left for Saskatoon. Expecting criticism for my garb, Sarsfield surprised me as they prepared to leave when he quietly said to me, "I like your outfit. Makes sense in this crowd."

This was the only meeting a superior of mine ever attended, even following my farm field rep appointment. I felt this was a vote of confidence in my abilities to handle the farm electrification duties.

* * *

AFTER WE HAD purchased the town of Kamsack electrical system our lines went north to St. Philips Mission and on to the town of Pelly. The little, isolated power plant there was run by the Ford car dealer and garage and a line went south to also serve the mission, including a general store, gas station, Catholic church and rectory, Indian residential school, the Federal Department of Indian Affairs agent's office and a few homes of white folks. A Mr. Davis was the Indian agent, brother of the *Kamsack Times* newspaper editor. He oversaw three Indian reservations: Cote, Keeseekoose, and The Kee. The homes of the Indian people would continue to be denied electricity. In fact at that time the constraints of the Indian Act, probably the darkest piece of jurisprudence ever enacted, still restricted these people to stay on the reserves.

In our preparations to serve the mission I had to meet with the folks in the settlement we would serve to have contracts signed for their accounts, including Davis for those accounts for which the federal government was responsible. When I entered his office about half a dozen native people were waiting to see him, but he rudely brushed them aside to talk with me. I said, "No, no, please, I'll wait until you deal with these folks who are ahead of me."

Belittling them with a mean snarl, he said, "Oh, to hell

with them. They're not important. They have nothing else to do but bloody-well wait." As we were about to finish the business I had come for, he asked, "Could the federal government buy additional poles for lines from the office to other departmental buildings?" I replied, "Yes, and we could dig the holes and set the poles for a small fee – extra 30 foot poles cost $14 each, $17 for a 35 footer, plus $5 per pole if you want SPC crews to dig the holes and set them." He snarled back, "I'll get these lazy, no good, ne're-do-wells to do a little work for a change. Just deliver half a dozen poles." Startled, sickened and saddened, I had never witnessed such overt verbal inhumane comments and public discrimination towards Indians. Davis, a duplicitous federal government Indian agent, whose job it was to help native people, was a major player in their cultural annihilation. He was diligently extinguishing any self-respect and esteem they might have had and, with it, their aboriginal traditions going back thousands of years of living in harmony and peace with nature and their deep respect for the sacredness of the land. My, how I disliked him! Their look of hatred towards him imprinted on me for a long time. How could you not have empathy for these First Nations people, so publicly oppressed, impoverished and degraded by a discriminatory, arrogant government and their abhorrent employee?

* * *

AT THE CONCLUSION of every organizing meeting I tried to sign-up one or, if possible, more of the farm committee so they would have examples in their possession of how the forms were to be completed. It was particularly important that each application show the exact farm yard location on its quarter section, needed for power line routing decisions and for estimating construction material. Then security bonds were completed for each committee person as protection of the funds they would collect on our behalf. My final words to them was a "give 'em snoose" pep talk, intended to instil a measure of enthusiasm. They were invited to call or write me about questions they faced for which they needed answers. The group was assured I would return and meet them again in about a month.

At one meeting the committee wanted every farmer who attended to sign-up that night and asked everyone to wait. Most did. So I sat at the school teacher's desk, completed over 30 applications which left the committee but a very few to get later from those who were absent. It was after two in the morning before we finished. They completed their work in a few days, not in a record time for a project sign-up, but they were resourceful.

* * *

WHEN A PROJECT sign-up was completed our engineering team prepared a work order which contained a list of the farms, their land locations and addresses, a detailed map of the specific line routes to serve all the new customers and a list of the needed materials. Head Office approval of the work order then prompted a series of actions. Each signed-up farmer was sent a "Wiring Notice," that they should plan to get their buildings wired as the lines would be built in the coming months. Those who had taken loans for the construction charges had their accounts set up so the repayments would be coordinated with their monthly energy accounts after they received service. The project was added to the construction schedule. The material; poles, conductor, insulators, transformers, meters, anchors and other hardware were ordered from various suppliers for delivery to the nearest railway siding or the corner of a farmer's field within the project or some other convenient location.

It was close to inevitable that after the sign-up was finished, the work orders issued and the construction material started to arrive on-site, that we would receive a plaintive plea from a previous hold-out to be added to the project. While it would have made economic sense to serve that farm the same time as the rest of the group, they were always turned down because we and the canvassers had made this commitment repeatedly and we were not about to break faith with our committees. Those latecomers were served the following year as a "single."

When the work orders left Yorkton for Head Office approval there was still one task remaining for both the farm committee members and field reps. It was getting free permission to cross private property where the rural power lines were to be built. More tales will be told later about the objections we encountered in getting that approval.

Self-Help

TED DURNIN, engineer in charge of construction, called our office one morning in the early fall of 1954, and asked to meet me that afternoon. He was concerned about the likelihood that many Yorkton Area projects could not be built that fall in spite of our commitments to the farmers to the contrary. I was devastated with the news as it was Garnet Parcher and me that had given many of those farmers assurance they would have service that year. There had been unusually heavy rains and millions of acres in the south-east part of the our area were flooded, like giant lakes, where the water table was above or just below ground level. These conditions prevented movement of heavy construction equipment on the soggy farmland until after the ground froze sufficiently to support their weight. If an early winter was in the offing with lots of snow the power lines might not get built at all. He told me, "Dave, we have to find another alternative to build the lines other than with our regular construction forces."

In Durnin's Yorkton Hotel room that afternoon he grimly reiterated his concerns and then exposed me to the concept of "self-help." In Cass-Beggs's 1947 rural electrification report he had mentioned the notion as an economic and practical way to build the massive amount of rural power lines over the long distances between farms, an idea that would not only get the work done but keep money in the community. The scheme had also been mentioned during the political orations in 1949 in the Legislature when the REA was being debated prior to its proclamation.

The concept was to have farmers use their hand tools and equipment, tractors and trailers to haul the wooden poles to each staked location, dig holes for the poles, set them upright, then tamp them in place, for which they would be paid the same as it cost the SPC construction crews to do this work. Once the poles were "planted" our wire crew would string the conductor and finish the job. This would keep SPC heavy equipment off the wet land. Several projects could be underway simultaneously. Durnin wanted to meet a farm committee to explore the idea in greater detail.

A project of about 30 farms, "Saltcoats West," was the closest to the city that was in this predicament. I called its chairman, John Baines, and arranged a meeting with his sign-up committee that evening at the Hilltop curling rink, on #9 Highway about 10 miles south of Yorkton. When I gave Durnin instructions how to get there, just as I was about to leave to go home for dinner, he emphatically said, "You're coming too! I need someone on my side who the farmers know and respect." We ate and headed to the rink.

Durnin explained the problem and the proposal to a skeptical Baines and his committee-men in the rink's waiting room. The discussion rambled not quite as orderly as the following outline intimates, but went something like this:

SPC would:

• arrange for the brush, trees, undergrowth and other obstructions in the power line right-of-way to be cleared. Bush-cutting was usually done with bulldozers, sometimes by farmers and, when it was, they were paid an hourly rate that was separate from what the rest of the self-help crew was paid.

• surveyors, stakers we called them, would mark each pole's location, showing its height and where the anchors were to be dug in. Normally there were 14 poles per mile of 13.8 kV rural power line, an average of 385 feet apart. Poles were 30 or 35 feet long, to keep the conductor high enough above ground for safe, unrestricted access of farm machinery underneath.

• train the crew of farmers to safely do their parts of the construction work.

• supply the self-help crew with the specialty tools; hole-digging spoons and bars, pike-poles to support the poles until they were tamped in place, and long wood bits to drill holes for the galvanized bolts to go through the poles.

• after the poles and anchors were in place an SPC wire crew would string the conductor and tie it onto the insulators, tighten anchors and install the transformers in each farmyard. Then the local DO would install the meter on the yard pole's circuit breaker and connect up the farmers wiring.

The farmer self-help crew would:

• recruit 15 to 20 men for about 10 days' work, appoint a foreman who would be their supervisor who would record each man's work and be SPC's contact person.

• load the wooden power poles of varying lengths on farm wagons at the material pick up point and haul them with their tractors to each staked pole location.

• drill two holes in the poles for bolts to hold the insulator pin, install the porcelain insulator, drill additional holes for anchor eye bolts and the transformer hangers in each farm yard. This part of the job was known as "framing."

• dig holes in the ground for the poles with a spoon and bar, each one to be 18 inches wide, five and a half feet deep.

• raise the poles with a farm tractor mounted front-end-loader, support and align the pole perpendicularly with pike-poles, then fill in the loose earth around the pole and securely tamp it in place.

• dig in the anchors where they were specified, as well as the grounding system at the base of the transformer pole in each farm yard.

The farm crew would be paid a flat amount of $6.50 for each pole hauled, framed, dug, set and tamped. It was suggested this amount be apportioned amongst the crew, dependent on the equipment used and the time required for situations which needed extra pay, like hole diggers or anchor installers who encountered large rocks and spent an inordinate amount of time to overcome those obstacles, or where poles in sloughs had to be mounded with stones to keep them in place, or to repair farmer owned equipment which broke down. Also, their foreman should be paid, so it was suggested the final apportionment be done after completion of the work. Additional amounts would be paid to dig in the anchors and the transformer grounding systems.

A 15-man crew for the Saltcoats 30 farm project would pay them about $3,500 for ten days work, an average of around $225 each. A farmer and his son could earn about what they had paid to get service, or help cover the cost to wire their buildings or purchase electrical appliances and equipment. For comparison my pay that year grossed $355 per month.

There were questions and doubts in the minds of Baines and his committee, but Durnin's persuasiveness, displayed wisdom, sincerity and reassuring manner convinced them that with their innovativeness they would be successful. They agreed to try because it was a challenge, a new idea and what farmer would not respond to something like this? And besides, what were the alternatives? They would be guinea pigs, the first project to give it a whirl.

We returned to Yorkton to start organizing the venture. Durnin called Bill Montgomery, a long-time project foreman who reported to him, to meet with us the next morning. Montgomery was skeptical of the idea, I think because he felt it made the work he did look easy if farmers could do it, but he respected his boss. Montgomery was asked, "Do you have someone on one of your crews, an experienced staker who can supervise rural power line construction; someone with the ability to safely train inexperienced workmen; is knowledgeable about the farm community; and able to get along with farmers and earn their respect?" Montgomery's response was almost immediate, "Ted, I think I have the right man on one of my crews: Glen Thorpe. But there's a little problem. He's back with a crew in Humboldt and I'll have to arrange for a replacement and then go and get him." Durnin asked, "Could have Thorpe back here tomorrow." Montgomery thought he could and left.

Next morning Montgomery was back with Thorpe in tow. Glen had been raised on a farm, was a big, burly man, over six feet, usually quiet spoken with a good sense of humour. His size alone commanded respect. The wheels started to turn quickly. Durnin confirmed the scheme's details to Thorpe with which Montgomery had already familiarized him. Glen accepted the challenge enthusiastically. Durnin had prepared lists of needed tools and had ordered many of them by phone from suppliers through the purchasing department in Regina. The three of them discussed how the training should be done and the formation a special SPC construction crew to do the staking, then string the conductor and finish the pole-line work once the poles had been planted. I supplied a list of projects where conditions of water in the fields dictated self-help as being the solution.

* * *

MONTGOMERY AND I attended Thorpe's first two self-help meetings with farmer-manned construction crews after which he was on his own. He trained and demonstrated the work the farmers would do and how to do so safely. He did a remarkable job and worked hard. Soon "Saltcoats West" and several more were underway, the lines were staked and the farmer pole crews started their work. SPC's 1954 annual report stated 200 farms had been served with electricity that year under self-help, "This method, used experimentally because of difficult conditions, was found useful and may be continued in future years." It did continue: a Made-in-Saskatchewan, practical and useful solution to solve an awkward and unpredictable problem.

Many lessons were learned that fall. Wooden-wheeled wagons could not be used to haul poles. Their weight was just too heavy because of their pressure treatment with creosote. The wooden hubs and axles burned out, so only rubber-tired wagons with roller bearings were suggested. Only the highest of front end loaders on farm tractors could safely raise the poles into the open holes, particularly higher ones when they were specified. Pole hole diggers who became adept with a spoon and bar made the most money. Poles set and tamped by farm crews were praised by the SPC wire crew as they felt safer to climb. They said the workmanship was better than when done with SPC employees. After all, those holes and the poles would serve the very people who put them there.

* * *

THE "HAMPTON-DONWELL" project of about 40 farms was self-help built in 1955 and included the two small hamlets. Its committee, like many, was skeptical about the scheme, but agreed to try. I had mentioned at the organizing meeting that a good pole-hole digger could complete two an hour. They insisted that I demonstrate, to put-up or shut-up. Before the farmer-crew training session Thorpe taught me how to use a spoon and bar in the yard of his Yorkton construction office. I dug one hole, but the ground was hard and stony, and it took me a lot more than 30 minutes.

A few days later we met the farm crew at the power line tap-off point on the east side of #9 Highway a few miles north of Yorkton. They watched suspiciously as I helped one of the diggers at the first stake and started the hole with a spade, got down about two and a half feet with ease. The earth was soft as we were in low lying, moist land. Then with the spoon and bar my cohort finished the hole. Thorpe left with the rest of the crew for their training. We moved to the second stake and repeated the process under intense scrutiny, finished in just under an hour. Convinced, the hole diggers struck off on their own.

I observed another training session at the rural line tap-off from a 25 kV distribution line. The rural line's route crossed a large slough and the first three pole stakes were in water. The farmers lightheartedly joshed each other how Thorpe would dig those holes. In his quiet manner he sent them to gather a wagon load of stones which they brought to the edge of the slough. Two power poles were lashed together as a makeshift raft on which the rocks were piled. He sternly told the diggers and pole-setters to follow him with their tools into the water as he towed the loaded raft behind him. The water soon came to Thorpe's thighs. The on-shore laughter abruptly ceased and they reluctantly followed, fully clothed. Before long, with a digging spoon,

Thorpe made a depression in the muddy bottom for the wood pole's butt, a pole was floated out, raised with some difficulty with the pikes and, when erect, the stones, with more rafted in, were pyramided around its base. When complete and the pike-poles removed Thorpe looked at them and declared, "There, that's how it's done," then left them to deal similarly with the other two poles. They earned their bucks that day. Those in the slough and the dry on-shore observers were soberly impressed.

Fun In The Slough[2]

I worked for Saskatchewan Power
Where some say there's little to do.
I answer with a challenge
Try digging a hole in a slough.

In my hay day as a lineman
When rain had flooded the land
You couldn't set poles with machinery
It had to be done by hand.

Standing waste deep in an icy slough
Soon sorted the boys from the men.
Dig for a while till it all caved in
Then start all over again.

Finally the hole was ready
Further across than deep.
Float in a pole and set it by hand
Then carry in rocks by the heap.

Right to this day when I hear someone say
You guys had little to do.
I simply excuse him as one of those boys
We sorted out back in the slough.

---

[2] From *Living Skies and Saskatoon Pies – Stories and Rhymes of bygone times* – by Claude Finch, a former SPC employee.

* * *

THAT WINTER of 1954-55 self-help was encouraged on every project I organized that would be built the following year. I explained the benefits but did not press for a decision at the time Their questions and deportment about the scheme helped me to glean their support when I met them after the sign-up was finished.

I used self-help as a key sign-up marketing advantage. A few of my jobs did not buy into the scheme, but in 1955 over 1,200 farms were served with farmers doing the construction work, nearly all of which I organized. A couple of projects who chose not to participate were content when the farm crew from a nearby previously completed job moved in and did the work and earned the money. I convinced over 1,500 farmers into self-help the next year.

For reasons not clear to me, then or now, the other area farm field reps did not take advantage of this scheme as much as me. Even though Harry Jessop and I shared an office, very few of his projects were built this way. It was almost exclusively a Dave Anderson initiative. Clint White quite incorrectly claims in his book, *Power for a Province*". . . the idea did not really catch on." I believe his comment was influenced by the relatively small number of 200 farms that were served in 1954, the first year the scheme was launched and mentioned in that year's SPC annual report. Self-help sure caught on with the farmers I convinced. In the last three years in which I was involved in farm electrification work, self-help served nearly 4,000 farms, more than 17 per cent of the total electrified in those three years, a number not to be sneezed at.

Self-help caught on in a big way with small acreage farmers or those facing cash-flow problems, the kind that were dominant in the Yorkton Area. The opportunity of earning a few hundred dollars and, to-boot, having the pride and satisfaction of building their own lines, took the sting out of costs that might otherwise have inhibited their signing up for service. Perhaps this explains the popularity of my pushing the self-help scheme for them "to get the lights."

Self-help, over its life, had a good safety record as there were no serious accidents ever reported from any of the dozens of crews. This was a credit to Glen Thorpe and his crew foremen that included John Denys, Erwin Kohlert and Jim Gilchrist amongst others, who went to have long careers with SPC. Later, Glen owned a power line construction business and contracted work with the Corporation and other utilities. In the 1980s he was appointed by the Devine Conservative government to serve on SPC's board of directors.

* * *

Saskatchewan – The History of a Province, by (Jim) F.C. Wright, was published in 1955 to commemorate the province's Golden Jubilee. It surprised me the author gave only a passing reference to the significant role electricity and natural gas was having on the province's economy and its citizens at the time. The impact of rural electrification on such a large and important segment of the Wheat Province's population is given but a mere two sentences on the book's fourth from last page. When this history was being researched and written farm electrification was in full swing with its huge investments and the spin-off benefits to the private sector to cope with that demand. The natural gas system was expanding with a customer acceptance level in every community that was served of almost 100 per cent. Wright gave the CCF's short-lived dabbling in a fish plant and a boot factory greater prominence. As one involved in the early development of both energy systems, I believe this omission was a major deficiency by the author about what had been accomplished in the province's first 50 years since its 1905 incorporation. Gee whiz, the story about self-help, a unique Saskatchewan development, by the book's standards deserved at least a sentence or two.

On the Road

ON THE ROAD in Saskatchewan, travellers on the prairies experience an unusual exposé of unparalleled beauty, an endless variety of countrysides and an overwhelming vaulted sky in this unique part of the Canadian plains. But this array of diverse, panoramic scenery is missed by all but a very few out-of-province folks and overlooked, even disregarded, by some locals growing up in its midst. They succumb to the cliche, it's so flat it's depressing and, with blinders on and pedal-to-the-metal, hurtle east or west on the Trans Canada Highway, then poke fun about its drabness and plainness when they arrive at their destinations. My hundreds of thousands of miles over that same landscape were anything but drab or plain.

On the road, I was stimulated and fascinated as one season melded into the next, with distinct contrasts between beauty and harshness, and each part of the province so different from the others. The south-west's rolling grasslands without dominating landmarks and covered with prairie wool had a special ambience where it was easy to imagine a lookout: a Cree or Assiniboine or Blackfoot or Sioux, on a knoll scouting for food or the enemy, where tens of thousands of buffalo once freely roamed, their thundering stampedes shaking the prairie for miles. The bald-headed Regina plains with their flat horizons where distances seem without limit. The prairie rolls into the north's Parkland belt of bluffs of protective bush and further on, dense forests flanking the hills covered with poplar, birch, pine and fir. Crossing the Qu'Appelle Valley was always a treat with its unique hog-backs running down to the valley floor to a string of delightful lakes connected by the meandering Qu'Appelle River with refreshing greenery along its banks. To experience nature on a rampage was exhilarating, sometimes frightening: hordes of ravenous grasshoppers, winter blizzards blasting with a vengeance, choking summer dust storms and nothing so ominous in this Big Sky country as a cloud build-up before a storm with howling wind, spectacular lightning crashes, terrifying thunder claps, then the welcome rain and sometimes devastating hailstorms. The haunting morning and evening yowls of a coyote, maybe a duet or a trio, sometimes a chorus, that sent shivers up your spine. Deer, antelope, moose, elk and other wild creatures running in countrysides where there were wide open spaces to run in. I drank in and was intoxicated by the colours of hundreds of magnificent sunsets and mesmerizing sunrises, unparalleled anywhere. In all seasons there were times when this outdoor environment was quiet and still, frighteningly so.

On the road in winter the stark bleak countryside was in a deep-freeze of quiet slumber for five months from Hallowe'en to April Fool's Day, a sobering time demanding respect particularly when it was very cold, accompanied with lashing winds. The sun kept bankers hours and often had visitors on both sides, intriguing sun dogs that signalled it was very cold and, at night, the Northern Lights danced on stage in spectacular grandeur with a supporting cast of stars: the Dippers, Orion, North Star and faraway galaxies, all twinkling on time and in their precise places. The air was clear, crisp and nippy on a 40 below morning when everything crackled and bristled. Yellow stubble protruded through the crusted snow in contrast to a neighbouring field of dark summer-fallow. Hoar frost covered everything some mornings – fences, grasses, trees, weeds, even the power and telephone lines had an aesthetic quality in their white shrouds. Dark brown sentinels of dock and other weed stalks stood at attention in roadside ditches, cattails were anchored in frozen-over sloughs. Leafless, barren trees and nondescript brown grass edged the roadsides looking very dead, but always fooled you and came to life after the snow melted in April. Machinery sat abandoned in fields where it last stopped. A weathered row of granaries and tattered, deserted buildings were stark reminders of the contrast between the land's appearance of infertility and its abundant richness. Some winters had more snow than others, but when a blizzard blew for days on end it all was on the move to somewhere else. Warm chinooks blessed the western part of the province and shortened their winters.

On the road in springtime exuded a special charm. Warm, dry winds licked-up the snow. April showers sluiced away winter's grime then all the greens arrived, giving new meaning to "a breath of fresh air," and with it, a promise of another season of optimism and growth. It was seeding time and machinery started to make the rounds and farmers proudly ensured their lines were straight for scrutinizing neighbours going by in their pickups. Migrant birds joined the stayers. Meadow larks practised their intricate scales, interspersed with a myriad of other birdsong. Canada geese honked their way by in V-formation fly-pasts. Waterfowl surfed to landings in runoff-full ditches and sloughs, where mating mallards had already taken up residence. Hawks, owls and falcons gliding over the fields or perched in trees seeking out their lunch. Bulrushes pushed their way up, saw their reflections in the water and stayed. Prairie bluebirds accepted hospitality in houses on fence posts, placed there by bird lovers. After the pussy willows came out, wild flowers emerged in delicate, pastel arrays. The earliest were dusty blue crocus, followed by blue beards-tongue, scarlet mallow, buttercups amid the purply-grey sage and prickly cactus nestled amongst rocks splattered with blue-green, black or rust coloured lichen, a signal they had been undisturbed for generations. In the bush the chokecherry and saskatoon bushes bloomed. Trees discarded their winter costumes of grey skeletons and readied themselves to be grateful shade providers. Gopher parents emerged on the dirt mounds around their burrows for protective peeks at their playful newborn and to warn them of foxes, hawks, owls and other raptors and predators. Drab roadsides turned to brilliant shades of emerald green. Jack rabbits shed their white camouflage and antelopes bounded stiff-legged over fields and fences. Deer birthed their Bambis in the bush and emerged in the open to feed on new grass. Storm windows came off schoolhouses. Farmers fixed fences, cleared land of stones and bush. Graders came out of hibernation to smooth the washboard and fill the pot holes on the gravel and dirt roads.

On the road in summer there was dazzling sunshine, its relentless heat making mirages magically dance over the shimmering plains. White, fleecy clouds scudded across endless blue skies. A full palette of crop colours, visual treats of green, gold, amber, yellow and blue fields contrasted with black summer-fallow. Shifting waves of growing and ripening grain, a never-ending symphony of undulating swells, stirred by the wind, interrupted only by a barn, granary, bluffs of bush, a farm yard shelter-belt. Fragrant lilac bushes of blue, purple or white graced the front door of almost every farm house or were a windbreak row for one side of the garden. And in the grasslands was the unmistakable smell of sage brush. Red-winged blackbirds chattered their tunes, clinging to the sides of cattails in dying sloughs. Sunflowers, daisies, goldenrod, prairie lilies, wild roses, alfalfa clumps, brown dock, white and yellow clover and yellow mustard painted the roadside ditches. Loaded saskatoon and chokecherry bushes invited pickers to gather the berries for their pies or puddings or pancakes. Hordes of land gulls followed a cultivator looking for worms and grubs in the freshly overturned soil. The smell of fresh mown hay was an odour never to be forgotten. With the summer came visits with a man of the soil in his field beside a slough where swarms of mosquitoes hatched hourly and were so thick they sounded like truck tires humming on hot tar. In some years crops were devoured by locusts and grasshoppers and they, along with moths, mosquitoes, butterflies and bugs, messed-up the windshield and plugged radiators that needed blow-jobs, as if the car had run headlong into into a pailful of scrambled eggs. In dry years strong winds spread dust storms that plagued the prairies, reducing visibility to only a few feet. On hot days when the wind was only a breeze miniature tornadoes graced the yellowing, sun-beaten countryside, "Little whirlwinds," we called them as kids. They picked up and carried skyward loose straw and dirt in their short, twisting life span before they died out and were replaced with others nearby. One could sense a farmer's intense pride and love of the land with well kept buildings and fields where weed control around poles, granaries and fences was neat and clean. Empathy and sadness pervaded one's thoughts when a hailstorm flattened cropped fields, and with it a farmer's hopes and dreams.

On the road in autumn spawned a sense of urgency as swathers patterned the ripened grain ready to be gobbled-up by combines with the hope fall rains or snow would hold off until it was all under cover. Some farmers hung on to the old ways and harvested with threshing outfits, the grain cut into sheaves, then manually stooked to ripen, waiting for horse-drawn racks and the crew of men on the threshers spewing grain into waiting wagon boxes and golden straw onto swelling straw-stacks. Along with the gold and khaki dress of the countryside was a unique, earthy smell as the machines reaped the earth's rewards, leaving behind the tell-tale plumes of dust and windrows of chaff and straw, or it was gathered up and baled into blocks, a sure sign the country was basking in Indian Summer. It was a time of thanksgiving with full granaries or even grain piled on the ground at some risk until it could be sold. The wind rustled the sandy coloured grasses and burst open the cattails with their cotton wool. At night came the spectacle of fires dotting the horizon as straw was burned in fields in preparation for next year's cultivation. As if as appearing from nowhere, crickets chirped their messages that fall was here and winter was just around the corner. Millions of migrant waterfowl headed south and rested overnight on lakes and sloughs or scoured for grain spilled from the harvesters in stubble fields. Breezes washed through the golden poplar leaves and shimmered them magically before they blanketed the ground. October precipitation came as razors of cold rain that turned to sleet, then snow as darkness overtook the plains, announcing winter's arrival. These first whiffs of snow skidded across roads to pile up along the fences and tall weeds. Coffee shop cronies speculated whether the snow that came before

Hallowe'en would still be on the ground when April rolled around, a sign of an oncoming tough winter.

On the road I was inextricably bonded to the land that is the Wheat Province. It was tattooed under my skin, always there, like being a Saskatchewan Roughriders football fan. I was part of next-year-country, that maybe-we'll-get-a-crop-this-year, or the truism of prairie agriculture, "Riskan Hope," emblazoned in large white letters on a black barn roof near Craik on #11 Highway between Regina and Saskatoon. It takes risk and hope and sacrifice, laced with a little luck, to grow a farm, a community, a province and a nation.

On the road had two downers for me. First, the incessant wind that I hated. Amid all this splendour it was always there, from a whispering, gentle, caressing lover in the evenings to a relentless and implacable noisy foe in daytime blowing dust or snow with grim determination that sent Russian thistle bounding across the countrysides looking for new lodgings. Second, I disliked the road-kill of nature's wild creatures under my car's wheels, on its bumper, windshield or radiator grille: gophers, snakes, coyotes, foxes, skunks, badgers, butterflies and a myriad of bird species. I tried to avoid them, but failed often, and in the case of skunks, regretted it longer afterwards than I would have preferred. Deer, or "jumpers" they were called, had many close calls with my fenders but always jumped safely in time.

On the road meant personal sacrifices, but this was neither demanded nor forced on me. Absent from my young family day after day, week after week, Jean more than ably raised our children almost like a single parent. They grew up with a part-time father as I was purpose-drenched in a job that was hardly work for me. I missed much of their childhood, a first step, a birthday party, bath tub fun, a first spoken word, loss of the first tooth, the first day of school, facets of family life that farm field rep pay scales failed to recognize.

* * *

ON THE ROAD, monotony from repeatedly driving the same highways and country roads could be hazardous, their twists and turns becoming embedded memories. Dozing at the wheel was an ongoing peril. With the first nodding hint I'd shake my head vigorously but if drowsiness persisted I pulled over, got out, maybe had a pee in the ditch and walked around the car. Fresh air from a rolled down window in cold weather eked out a few more miles. I was jolted awake a few times when the tires brushed a gravel ridge on the highway's edge or slumped off the roadway as the car headed for the ditch. Usually I just pulled off the road, slouched down, laid my head on the back of the seat and had a short nap. My official company safety record listed no accidents but there were several close calls.

On the road, potential accidents at night were heightened after the 10 o'clock beer parlour closures as plastered locals haphazardly zig-zagged down the highways on their way home. Time spent learning defensive driving tactics from the province's Traffic Safety Branch was time well spent. Another danger, although only rarely encountered, was farmers on tractors or combines, weary from long hours in the field, carelessly coming onto or crossing provincial highways without stopping or checking traffic in both directions. Twice I hit the ditch for survival and there were other not-so-near misses.

On the road, searching for a specific farmer's yard with directions on how to get there was easy, even in the dark hours, not quite as easy as a Ukrainian cohort once said, "Just drive the road down," but easy, like, "Three miles west of town on the highway, turn north four miles, then three and a quarter west, our red barn is on the south side of the road." The province's land surveys made the roads all north-south and east-west. The prairie sun was a big help year-round to keep one's orientation when in the country. But when it was cloudy or during a storm more attention had to be given to one's whereabouts. Our car odometers were invaluable instruments. The metric system of kilometres instead of miles has doubtlessly since complicated this, as it has for older farmers who now have to think about litres rather than gallons, grain in tonnes instead of bushels, acres are now hectares and there is a need for two sets of tools, metric and imperial, for machinery repairs.

On the road, slowing down or pulled-over and stopped to allow passage of a farm tractor with wide equipment in tow, or a team hauling a hay rack bulging its mushroom-like load, their drivers always gave an appreciative wave as they passed by. Conversely, whenever I pulled over and stopped to check a power line route map detail, look at sign-up papers or search a file for a farmer's land location, a passing farmer in his half-ton or on his tractor always stopped to check that I was all right or to offer directions in case I was lost. These brief encounters gave me a good feeling to be in their neighbourhoods.

* * *

ON THE ROAD, socializing with other commercial travellers soon exposed me to the advantages of membership in the United Commercial Travellers and the Associated Canadian Travellers associations of cheaper hotel rooms, a benefit to SPC, and of benefit for me were their group life and accident insurance plans that gave coverage for personal injury while travelling, a fringe benefit the Corporation had yet to embrace. Certainly the number of miles I drove far exceeded the two associations required minimums. However, my applications, supported by my soap, meat and grocery salesmen counterparts, were always turned down as SPC was considered part of government and public servants were unfairly excluded.

* * *

ON THE ROAD without credit cards, before their invention, meant carrying enough cash for a week's supply of hotel rooms, meals, gasoline, engine oil, flat tire patches and other minor car repairs. Cash advances enabled use of SPC funds instead of our own. Major vehicle maintenance such as tires and motor or transmission overhauls were covered with field purchase orders and issued to garages and service stations which then billed Head

Office for payment. Receipts were needed to claim hotel room expenses and only when they exceeded the limits for meals, but it was wise to not tempt the hawk-eyes in accounting as it seemed they had memories like elephants if there were too many gourmet meals or overnights in deluxe bridal suites.

* * *

ON THE ROAD in the 1950s, dingy hotels were my normal overnight lodgings. Rates varied from $1.25 to $4 a night. Motels, few and far between, mainly in the cities, were more costly, like $5 or more, and a luxury beyond our means. The spartan rooms I inhabited night after night were generally clean but were shabby in the smaller centres, Rose Valley's being a pleasant exception, as compared to those in larger towns like Melfort, Tisdale, Nipawin or Humboldt. Even in these larger centres they were pretty plain pads. All were a far cry from the standard, comfort and lifestyle of our homes.

On the road, small-town, clapboard flop houses, were usually outmoded buildings. Many were fire traps. The rooms were kept open only because provincial liquor licensing regulations required their availability for travellers if the owner wanted the more lucrative ground floor beer parlour. These parlours were male-only dens of iniquity until 1962 when women were permitted to enter and they were renamed the more classy "Licensed Premises." The last available rooms for rent in most hotels were those that were above those suds sanctums. The inevitable noise from down below meant little or no sleep until the joint closed and the revellers ended dimwitted discussions or senseless scuffles on the street below. Most rooms were without desks so paperwork was done by clearing off the wash stand or sitting on the bed using a briefcase as a table. Single beds, 39 inches wide, held thin mattresses on baggy, stretched springs. Many times, late at night, drugged with weariness, I plodded up the flight or two of stairs and flopped into these sacks and slept the sleep of the dead.

On the road in June 2002, visiting the Grasslands National Park in south-western Saskatchewan, near Val Marie, population of about 300, I stayed in the two-story Val Marie Hotel. I had called ahead for a reservation and asked, "I would like a no smoking room not above the bar." The owner's response, "We have no smoke-free rooms and all of them are above the bar." This should have alerted me. On arrival I was given one of their six rooms at the outrageous rate of $45 per night. I thought my eyes had betrayed me on opening its door. It was the spitting-image of those I had occupied 50 years earlier, 10 by 12 feet, a bare 60 watt light bulb suspended in a brass socket hung from the middle of the ceiling, a 54 inch mattress bed, a broken desk with the loose pieces sticking out of a drawer, one hard-back wooden chair, no television, no radio, no telephone, not a picture on the wall, no bedside table or lamp. The traditional wash stand, bowl and water pitcher was replaced with a sink that did have running water and above it a small mirror and a 60 watt bare bulb and an electric razor outlet. The water was hard as nails, soap lather a pipe dream. The shag rug was filthy, beyond anything I ever put my feet on, as was the flush toilet down the hall that didn't work some days. Outside my window hung a bar sign, "The Lucky Lady Licensed Dining Room." It squeaked in the never-ending wind as it swung to and fro. This element of rural Saskatchewan has apparently remained unchanged.

On the road, hotel fire exit locations were always implanted in my mind in the 50s after check-in. I sometimes even checked to see if those doors really opened. Rooms on the second and higher floors usually had a built-in do-it-yourself, emergency fire escape, a hefty rope knotted about every 18 inches, anchored near the window and long enough to reach the ground through an opened or broken window. There were guests, after lifting a couple too many in the beer parlour, who verified these escape routes. Some hotels did have fires, like Raymore, where the top floors were demolished and later rebuilt.

On the road without running water in hotel rooms was more than inconvenient and primitive. It was a hardship. Most municipal roads I travelled on were gravel, but many were just dirt trails, the earth simply dug up from the side ditches and levelled off. Some farmers gravelled roads from their farm yards to the upgraded municipal road so they could have all-weather access with the outside world. In 1953, the only pavement in our area was a few miles of #1 Highway east of Indian Head to Grenfell, #10 from Duff to Melville and on to Yorkton via Willowbrook, and #9 Highway from Yorkton to Canora. The rest of the provincial highway system in our area was gravel, so choking dust in our vehicles was routine. The panel vans and later, station wagons, drew in dust through the leaky back doors like a vacuum. At day's end it was impossible to get refreshed with a washcloth in the wash basin with a quart or two of cold water from a pitcher in which often floated a dead fly, moth or wayward ant. So the communal tub at the end of the hall, if there was one, shared with 20 or so other guests, was reluctantly used. Many bathrooms were just high-level pit toilets. Hot water to shave and wash in the morning was available providing, the night before, your room number on the front desk call sheet had been circled for the desired wake-up hour. At the prescribed time the night desk clerk knocked, awaited a response, then announced the time and left a jug of warm water on the floor outside the door.

On the road, hotel rooms in winter were, in some joints, like ice boxes, the heating system far from adequate. If the lobby stove was a supplemental heat source it was prudent to get a room near the staircase and leave its door open overnight. In summer, air conditioned rooms were unknown, but of course the beer parlours had them. On warm days a late check-in to a three-story hotel without a reservation meant the only available rooms were likely on the top floor, a guarantee for a hot, restless, uncomfortable night after the sun had beat on the roof all day. If the room faced north or west, the prevailing wind direction, maybe a cool breeze through the window made them more tolerable, another pleasant prairie phenomenon that even on the hottest days a cool evening could be anticipated.

On the road, lighting in hotel rooms was a single, naked bulb in a brass socket suspended from a cord in the centre of the ceiling, sometimes controlled by a wall switch. It was difficult to read, write reports, work on maps, check applications, receipts and loan forms as bulb wattages were usually 40s, maybe 60s. I often replaced these with higher wattage 100s, street light replacement lamps I obtained from DO's, and left them in the sockets of the rooms I occupied, a small attempt at building more load for the electric system.

On the road, fleabag hotels were exactly that occasionally, because of bedbug infestations. The grapevine quickly broadcast this news amongst regular travellers to steer-clear until an all-clear was sounded. When unsure if vermin might be alive and well in some establishments I checked the underside of mattresses before crawling in between the sheets.

On the road, motels were permissible near the end of my farm field rep days. I once got the last available room on a very cold night, an end unit that faced into the wind. The forced air heating system furnace, 10 units or more away, had little oomph left in the ducts by the time the hot air was supposed to reach my room. Unable to get warm in bed, I sat, wrapped in a blanket, with my feet in the bathtub's hot water until I warmed up enough to get into bed and to sleep. In spite of this experience motels with their individual bathrooms and desks with a table lamp on them were an appreciated luxury over the old hotels.

* * *

ON THE ROAD meal allotments were, as with our lodgings, adequate, but certainly not generous considering we worked hard, long hours needing three squares a day. T-bone steaks, pork or veal chops or fried chicken or apple pie a la mode were well beyond our means unless they happened to be that day's restaurant "special." These deals were the best value: liver and onions, roast beef, hamburger steak, vegetable and meat stew, veal cutlets, meat loaf, with desserts of jello or bread or rice or caramel or tapioca pudding for dessert, the latter one we called "fish eyes and glue." Often I skimped at lunchtime, then splurged at dinner to keep within my daily allotment. I often wanted to bring along those who guarded the company purse strings on a week's trip to see how well they would fare on this fare.

On the road small town cafes were always on Main Street, usually Chinese owned. It was wise to patronize the one with the lineup of pickups or tractors in front, a sure sign this was best place, the local yokel hangout, where discussion of the day's events was accompanied with their bottomless coffee mugs, smoking twisted-end roll-your-owns. Some had counters with stools, maybe booths, but most were simple chrome tables and chairs which, in summer, were festooned overhead with umpteen sticky flypaper twisted rolls to catch moths, mosquitoes, bugs and flies. The owners of these greasy spoons were often castoffs from railroad construction gangs who had dispersed across the prairies when those jobs needed them no longer. They went on to open eateries, plainly named Star, Gem, Sam's, Charlie's, Ace, Lee's, Wong's. These were decent folks, respected, polite, well-liked and active community citizens. Their restaurants usually lasted only until the parents could no longer work the long hours, then they closed when their offspring determinedly moved away knowing there was a better life than in what they had been raised.

On the road travelling for Steele Briggs in earlier days had taught me how to be road-smart and avoid the hazards of food poisoning from meals like hamburger steak or meat loaf if the town did not have electricity and the cafe lacked refrigeration equipment. In those cases I ordered bacon and eggs, or a bowl of canned soup, maybe a plain tomato sandwich in summer. Then in some eateries there were those awful dried-out roast beef or pork specials that needed a chisel to cut bite-sized pieces.

On the road exposed me to the home-cooking prowess of the wives of my farm committee members. Almost without exception I had no trouble leaving a whisker-clean plate after those treats. These hearty repasts were sincerely offered in appreciation of my efforts. Mid-morning or afternoon lunches in any cozy farm kitchen, bathed with a homey scent, were usually comprised of sumptuous homemade bread, canned jams and preserved jellies, maybe a jug of buttermilk, pie and cake. If there were children in the family I left them a dollar or two to spend the next time they went to town.

On the road my favourite eatery was the tiny New Canada Cafe in the little village of Kandahar, owned by an affable Chinese Canadian and his son, assisted by their wives. It was on the main street, a little ways off #14 Highway which paralleled the railway tracks that ran through the village. Cafe booths sat four patrons, plain wooden tables and benches with high side-wings which provided an element of seclusion. I came to be familiar enough with the owner-cook-waiter that it was routine, without asking, to get a shot of neat rye without ice, illegally served in a coffee mug after my order had been taken. Their specialty, that became known far and wide, was huge T-bones that hung over the plate's edge smothered in mushrooms, onions and deep-fried potatoes, cooked in steel pans over a wood fire. The owners were obliged to build a new cafe, to survive when the highway was rerouted around the south side of the village. The overall ambience did not match that of the earlier premises. It also reduced the size, tenderness and flavour of the steaks, then cooked, I have to admit, electrically.

On the road, meals and lunches were also a treat in Rusnell's Cafe in Wadena. It was high on my eatery list, a friendly place with close to home-cooked, not fancy, but flavourful food. Hockey posters, banners and pictures hung on the walls and reflected the interests of the father and his boys. Perhaps the most famous was Dave who was later a popular forward with the World Champion Trail Smokeaters.

On the road the Tisdale Hotel was a stopover for me whenever possible. Its rooms and dining room were well above average, clean and bright. The friendly husband and wife owners and their staff went out of their way to cater to us "regulars." We responded and went out of the way to overnight there. Evening dinner specials from Monday to Thursday were generous and tasty and, more importantly, thoughtfully within the expense account allowance.

On the road in Sturgis, about 1956, a lady set a few tables in her home's dining room. It was always packed with travellers who drove long ways to patronize her. The business boomed, so much so that it became too much for her, and its success from her hard work was the cause of its early demise. The meals she served were like those at a prairie fowl supper after harvest time. These were events that happened on Saturdays and Sundays, when farm field reps were at home and off the road, unable to take advantage of these popular Thanksgiving events.

On the road, Fridays were fish day in many restaurants, catering to those whose religious persuasion required them to get their proteins without meat. Often these meals came from cans: sardines on toast or salmon loaf or smoked finnan haddie. Some of these offerings weren't bad depending upon the dexterity of the cooks. Fresh fish was never available with the exception of a few cafes in the northern towns of our area, like Nipawin or Carrot River, where commercial fishermen's catches of whitefish, pickerel or jackfish from the lakes further north were dished-up.

On the road, breakfasts anywhere were relatively inexpensive, maybe a dollar and a half for bacon and eggs, toast and coffee, which tasted pretty much the same no matter how bad the facilities or cook. Even the smallest hotels usually had a few tables set for a breakfast of some sort.

* * *

ON THE ROAD was depressing in the fall of 1954 in the southern part of our area, as mentioned before, because the countryside was inundated from unseasonably heavy rains. Farmers could not harvest, if they had any crop left, until after freeze-up when the sodden ground froze and it could sustain the weight of their equipment. Our projects in this wide area had to be deferred for at least a year with the reluctant agreement of the farm committees after we assured them they would be first on our lists when local conditions improved. We shifted our program plans from the drowned-out areas to those blessed with good crops, mostly in the north.

On the road that winter I organized many projects on the northern fringe of our area, including several hundred farms north of the Saskatchewan River on both sides of #55 Highway which connected the communities of White Fox through Love, Garrick, Choiceland, Snowden, Smeaton, Shipman, Foxford and Weirdale. A 25 kVA three-wire distribution line followed the highway and was the power source for these projects. I worked for weeks, with Nipawin's Avenue Hotel as my operational base, and helped its owner, Jim Coulter, reduce his mortgage. He had renovated one wing of the hotel with modern room furniture and, more importantly, private bathrooms. As a regular customer he moved me to a new room two nights a week, if one was available. My weekly hotel bill showed only the average cost for the four nights, two of which were in the more deluxe units, the weekly room total within my expense account limit. I was grateful for this charitable gesture. Until the mid-60s SPC's living-out policies relegated us to roles of second class citizens as compared to other commercial travellers who enjoyed more liberal expense account allowances and could afford better meals and live in the more convenient and better accommodations offered by motels that started to spring up throughout our area.

* * *

ON THE ROAD, working with farmers meant I occasionally stayed in their homes overnight because of snow that blocked roads after I had arrived there to do my business. One adventure happened after an organizational meeting in the General Store in the hamlet of Welby, the first settlement a few miles west of the Manitoba-Saskatchewan border on the CNR mainline that runs between Winnipeg and Saskatoon. My "Welby South" farm committee chairman was Arthur Kelly, a Qu'Appelle Valley homesteader, a fine, well respected, elderly gentleman. It was a cold and stormy night. Snow was falling heavily and it had started to drift as I made my way to the store. The wind rose during the evening to almost blizzard conditions. This part of the province only gets about 16 inches of precipitation a year and I wondered why most of it had to come that night. I knew I had no chance to get back to my hotel in Esterhazy and this unsettled me as the evening wore on. When the meeting finished Kelly told me I should follow him, as we could probably still get to his farm where I was welcome to spend the night.

On the road, getting to the north rim of the Qu'Appelle Valley gave us no difficulty through the swirling snow. We crossed the valley floor, but travel on the preferred south-side road was impossible as it was already clogged. The north-side road was clear of drifts, being in the shelter of the adjacent hogback hills, and the snow that was piling up was manageable but was offset with another problem, treacherous, icy stretches of roadway that sloped towards its edge caused by water which trickled and oozed from the valley wall and froze, more on the hill or up-side of the road than on the down-side. At each icy section we cleared the snow and chopped a rut a bit wider than the car tires, about three or four inches deep on the down-side of the sloped road. An axe was not part of my survival gear, but Kelly had one, and being younger, I chopped under his supervision in the beam of the car headlights despite the wind and the snow which continued to fall and churn around us. I drove, one car at a time, under Kelly's directions as he walked ahead of the car guiding me to keep the left side tires in the rut so it would not slip off and into the ditch or over the embankment. It was an exhilarating experience for me, but Kelly had done it before and seemed comfortable with the process, at least that was my impression. The longest icy stretch over which we inched our way was perhaps 75 feet. Most were about 10 or 15.

On the road after about two hours of shovelling and chopping we arrived safely at his farm and warmed up in the kitchen. His wife had already retired. Their spare bedroom was above the kitchen, accessed by a pull-down ladder with one end hinged to the ceiling. Kelly pulled it down, gave me a flashlight, then gestured for me to clamber up and open the trap door. We said good night and with the door closed I found myself in a pitch-black, unheated, windowless attic, almost as cold as it was outside. Certain I would perish before morning I undressed and gingerly crawled into the bed. Its only cover was a feather tick that enveloped me and was surprisingly warm. I slept uninterrupted until their breakfast preparations woke me. Kelly's snowplow club had already been alerted to my need to get out and had cleared his municipal road west to #8 Highway. After breakfast I left for Esterhazy to retrieve my suitcase and pay for my unused hotel room. The storm had petered out. It was calm and the morning sky was blue and brisk and clear – a beautiful snappy cold day.

On the road was tough with record-breaking heavy snowfalls in the 1955-56 winter. The "Fosston West" farm group was in difficulty with their sign-up. Their chairman, Francis Eyre, appealed for my help. Their country roads were hopelessly blocked, even the Highway Department had officially closed #35 Highway from Wadena to Tisdale because numerous blizzards after New Year's had caused heavy drifts, some over 20 feet high, that abandoned any hope it would be opened before spring. Leaving my vehicle with Verne McQuarrie, the Wadena DO, I took the train about 10 miles north to Hendon, the closest railway station to Eyre's farm. He met me with his travel mode, a one-horse drawn stone-boat. Bundled in my survival gear and with briefcase and Gladstone wedged between my legs we held on to each other for support and bobbed over the huge snowbanks to his farm. For two days I was their house guest. From early morning to late at night we travelled on the stone-boat many miles the length and breadth of the project. We had breakfasts there and some dinners. Our other meals were at either homes of committee members or those we canvassed. More will be told later about this venture.

On the road in the early 60s I spent two nights in native homes, one was with the band chief, the other a band manager, and their families. I was storm stayed in both instances following band council meetings on the reserve. This was after the Federal Department of Indian Affairs had finally relented and decided electricity would be provided to indigenous people whom they had cruelly corralled into Indian reservations. I was always made as welcome and as comfortable there as in any other rural home.

The hospitality and geniality shown to me implanted a soft spot in my heart for these folks who were trying hard to lead their people to a better way of life. My hosts were not the stereotypes many whites in society relegate to these people as being lazy, dirty, drunk, immoral, savage and illogical. Some of them are, just as there are some white folks, but most Indians are like they are because the white man put them there.

* * *

ON THE ROAD one winter Friday morning in 1956, while heading for the farm committee chairman's home to finalize their sign-up details on the "Codette West" project, CBK radio broadcast a storm warning. I hurriedly finished my work and high-tailed it for #35 Highway, turned south and headed for home. I planned to try to outrun the storm going by way of Tisdale and Hudson Bay, a longer but more sheltered route than proceeding on south, then east to Yorkton. Before I had gone the first 30 miles to Tisdale the storm-front hit, strong north-west wind and heavy snowfall. Drifts started to accumulate on the road and as the wind strengthened, visibility deteriorated. To have continued would have been foolhardy. The storm was a dandy, mountains of snow and a blizzard that swirled for 48 hours, blocking all highways in and out of town. Home was the Tisdale Hotel for two days for three stranded Yorktonites: myself, soap salesman Bill Fleming and Nick Bretherton, a colourful retired RCMP officer who, 30 years earlier had, according to him, almost single-handedly maintained law and order in the area from Yorkton to the Manitoba border and north to Kamsack and Canora. He had also been a justice of the peace in this same area. He regaled us with many fascinating tales of how the west was won – great entertainment between games of cribbage, reading, sleeping, eating and waiting out our predictions about the storm. It was Sunday at noon before the wind abated. We heard a highway snow plow would head east to Hudson Bay so we checked out and were a convoy of three slowly being escorted home.

* * *

ON THE ROAD holding evening organizing meetings week-in and week-out meant our daytime activities were with nearby farm committees who had finished their sign-up canvass. I would check and gather-up the signed contracts, loan and application forms and several thousand dollars in cash, anywhere from $10,000 to $20,000, including many negotiable and endorsed produce vouchers and cheques issued by cattle buyers or creameries, along with non-negotiable cheques payable directly to SPC. Over the five or six winter months in the 50s I picked up around three quarters of a million dollars from the farm committees and safely shepherded it back to the office, without a cent ever being lost or stolen. It was not unusual to find in these stashes several original bank note bills from earlier times when individual banks issued their own currency. This was before the Bank of Canada printed and issued standardized notes, replacing those of the various banks. I often mused about exchanging those in mint condition for my own money to cash-in on their potential numismatics value, but never had enough personal funds to do so. This cash and the produce vouchers were converted into drafts payable to SPC at local banks or credit unions after they opened the next morning. But the bundle of moolah had to overnight in an envelope under my pillow in a room locked with only a skeleton key, a duplicate of which at least a dozen others had in that hotel those nights. It only happened once that I got up, shaved, dressed, breakfasted, packed and departed town with my briefcase and well-worn Gladstone, but not the money. Twenty-five miles down the road I remembered the forgotten envelope and its contents. Devastated and frightened out of my wits I white-knuckled it back to my hotel room. Its door was wide open, as I had left it, awaiting the chamber maids, the untouched envelope still safely out of sight under the pillow.

* * *

ON THE ROAD for our 1955 family summer holiday travelling to some far-away place was not an option with a young family. So we booked a cabin for a two-week July stay at the Kenosee Lake Lodge in Moose Mountain Provincial Park, north of Carlyle. My wife, three-year-old son, infant daughter and I arrived Friday evening, checked in and paid for our intended stay. The next morning on the beach, as we relaxed in the sand and sun, George Lincoln happened to walk by. He was reeve of the RM and had been our farm committee chairman the year before on the "Wawota" project. It was a large job which brought electricity to some 90 farms, the provincial park, the town of Wawota and the hamlets of Dumas and Kelso. He had been an exceptionally good chairman, a decent, solid, salt-of-the-earth type of citizen. Our acquaintanceship on the job had grown into friendship. After visiting on the beach he left only to return a short while later with some cash in his hand and a set of keys. He had had the balance of our cabin rental refunded and told us we had no alternative but to move out of our cabin and into his nearby, newly built cottage, as his guests. He took us there. We moved and did not see him again until the evening before we were to return home, when he came to see how we had enjoyed ourselves. This gesture of appreciation for my work with him and his committee had deprived him and his family the use of their new cottage for that period of time in Saskatchewan's short summer holiday season. Lincoln was a staunch Liberal party member, but that I worked for a CCF-run crown corporation made no difference in his show of gratitude. As time passed I visited him whenever I was down his way. We corresponded regularly for several years as he aged and moved into a nursing home.

* * *

ON THE ROAD in rural Saskatchewan had other rewards too. Gifts of food from wives of grateful farmers or committee members in appreciation for my work on their behalf, a jar of pickles or canned chicken, a sack of corn on the cob, a couple of dozen eggs, a dressed chicken, sometimes a turkey and once, a goose. If I happened to be on a farm after frost had blackened the vegetable tops and garden flowers and it was potato-digging time, a bag of spuds or a box of beets was waiting for me at the rear door or trunk of my car when I was about to leave. One Saturday winter afternoon while downtown in Yorkton shopping with my family a frozen turkey was left on the steps at our home's front door, its donor a mystery to this day.

On the road substituting for Garnet Parcher during his terminal illness, he asked that I stop at an Ituna project committee chairman's farm, a job he had finalized before falling ill, to pick up half a dressed deer, a gift for him and his family. These were not beforehand bribes in any sense of the word because they usually were offered after our sign-up work with the donors had been completed. Today, in this mixed-up world, an opinionated, meddlesome media or some prying two-bit politician would label these gifts as undeserved kick-backs or an under-the-table payoff deal that would have convicted us as being in a conflict of interest. Our farm field rep pay scales didn't recognize this benefit.

* * *

ON THE ROAD I met a lot of dogs, the four-legged variety. Almost every farm had at least one that inevitably announced the entry into its farmyard of any and all strangers. I liked most dogs, had had several in my younger years as did my children, and I respected their protective and security role on farms. Most seemed happy to see a stranger, their tails wagged friskily when I rolled down the car window to talk quietly to them after they dutifully had lifted their legs, trying to get my car's tires to sprout. Maybe they knew I liked dogs, a sense we humans don't yet fully comprehend? Others were more aggressive, barked and growled with neck hair bristling, wagging slow tails and looking like they'd relish a taste-test. By this time someone usually appeared from the house or barn or another out-building. If no one showed I usually got out of my car and, with an air of confidence, slowly made my way to the house, with the animal in plain view as I talked to them with my hand outstretched, palm-down. This routine worked as I was never bitten or attacked, even by the most aggressive ones, but must admit there were a few that were nasty, made me retreat ass-backwards or kept me in the safety of the car when no one was home or until the creature was in control by the farmer or someone in his family.

* * *

ON THE ROAD in rural countrysides meant going through gates. Some swung on hinges, and were made of lumber and kept freshly painted. They often announced the farm's name or that of the family, suggesting that maybe entry was by invitation only. Most though, were simple: three or four strands of snarly barbwire with loops at top and bottom of the end-post into which the gate-post was securely planted then hooked on top. A cardinal rule never overlooked, in spite of their being a nuisance, was if a gate was opened it was always closed after passing through. It was inconvenient having to stop, get out, open the gate, get in, pass through, get out, close the gate and then get back in and proceed only to repeat the process on the way out. I always said a little prayer of gratitude to the Texas gate inventor, a convenient time saver.

On the roadside barbwire or page-wire fences were often posted with No Trespassing or Keep Out signs, intimidating when calling on farmers who had refused to give us permission to cross their land with our power lines and might not be pleased with our presence. Usually the bark of the sign was worse than its bite. The farmers did mean business though when the signs read No Hunting, a pastime in which I was never involved.

* * *

ON THE ROAD, off beaten tracks sometimes meant farmers cultivated and seeded crop right through the ditches and onto road allowances, up to and over the sparingly used dirt trails, wanting to milk as much as possible from what was otherwise unused land. The grain stalks between the tire paths brushed against the car's front bumper, fenders, doors and the undersides. Close to harvest time when the grain was in head, going slow was critical to avoid breaking the stalks.

* * *

ON THE ROAD in panel vans and station wagons meant we drove vehicles stripped down of anything and everything that had the appearance of convenience or luxury, bulk purchased from the lowest tenderer and unsuited for the conditions under which we travelled. As mentioned earlier very few miles of the provincial highway system in the Yorkton Area were paved and driving 35 - 40,000 miles or more a year on gravel roads, many of the washboard variety, these rust-buckets literally fell apart. Very likely they were not bolted together right in the first place, well below their manufacturer's advertising claims. Some lemons lasted barely a year, should not have been let out of the garage. Dodges and Plymouths were notoriously inferior, one model not able to make it through one year of use. It annoyed me that in good times the Chrysler corporation, a bastion of free enterprise, a ferocious free-marketer, quickly flip-flopped in 1980 into what private enterprise usually refer to as simpering corporate socialists, when financial difficulties drove them, cap in hand, to government where they expected and got an undeserved bail-out from the public purse. The quality of the vehicles they produced did not deserve such treatment.

On the road in rattling Fords meant enduring drive-train related mechanical difficulties. The engines burned oil prematurely or burned out when driving through mud or snow, as did the transmissions. One of six new station wagons assigned to field reps, including me, didn't get 10 miles from the dealers lot. The engine burned out because its oil circulation holes were not drilled out at the time of its manufacture. The bodies on some were bummers. General Motors' touted slogan of quality "Body by Fisher," was a joke. Most vehicles clattered and jangled from their first mile to the last. They had doors that didn't fit, hoods that twisted in the wind, doors and windows that were drafty and leaked in dust like a sieve. The quality of engineering, design and manufacturing by North America's big three automakers have been, for generations I believe, clear cases of planned obsolescence, duping drivers and owners year after year. By the 1950s, metallurgy science was well advanced, certainly far enough along for them to have specified higher quality workmanship and materials in their products. They had been in business long enough to have been more ethical. They believed the transparent claims of their own advertising. It was not until the mid-60s, driven to higher levels of accountability prompted by consumer advocate Ralph Nader, and the superior quality of offshore competition, that improvements started to emerge. And even yet, as any honest Canadian Tire mechanic will tell you, Japanese made cars are still superior.

On the road saw my left arm dangling out the open side-window in the summer as it tanned several shades darker than the right, accompanied by dust and insects that streamed in the open windows spawning pipe-dreams of driving with an air conditioned interior.

On the road long before tubeless tires were concocted meant we drove with rubber inner tubes that could not stand up to the winter's freezing cold. I carried two spares and often during the week needed both. A great starter for the day was being greeted by a flat tire on a cold morning in front of the hotel. Jacking up the car and changing tires on the roadside in all weather conditions was routine, the flats caused by the frost breaks, sharp stones, nails or needle-like metal shards that had worn off the highway road-grader blades. The jacks we carried were of the bumper variety, standard equipment that came with our limousines. Their use was a hazardous manoeuvre if the car was on uneven ground because it had a propensity to slip off the jack, as it also did on a slanted roadside if the wheels had not been securely blocked with stones to prevent the car from rolling forward or back. I carried my own wind-up jack that worked better when used under the car's axles.

On the road in winter for me was before hotels had electrical outlets in parking lots to connect block heaters, battery blankets and interior car warmers and, when they did, there was usually an insufficient number for all guests, so to be sure to get one an early room check-in was critical on cold nights. I carried a long extension cord and sometimes ran it to an unused outlet or, if there weren't any, out of my room window. Jumper cables were essential umbilical cords to boost dead batteries. They helped others as much as they did me.

On the road in Nipawin one 40 below zero Friday morning my driver-side door window shattered into thousands of tiny, hexagonal pieces when it was closed. This was caused by a metal pressure point on the glass, a manufacturing defect. The local Ford dealer had no replacement so a piece of cardboard was crudely fitted and taped in its place. I headed south to Tisdale but that Ford dealer was out-of-stock too. It was a chilly 200 miles, the cardboard needing frequent retaping and fitting as I high-tailed it home to Yorkton.

On the road survival gear was in my car's trunk from October to April. Candles to burn in an open-top tin can for heat, wooden matches, a couple of tins of canned heat, two or three chocolate bars, a loaf of bread, two or three cans of gas line anti-freeze, a tow chain, a long-handled snow shovel and a couple of blankets. Extra clothes included wool socks and long-johns, zippered boots, leather mitts with wool liners, a toque, a balaclava, a wool scarf and a war surplus windproof, thigh length hooded parka. An old briefcase held a complete set of up-to-date RM maps for our area, for possible use in emergencies to check my location in relation to the nearest town or farm where I could safely trudge for help.

On the road snowbanks in winter weren't much fun. I was frequently stuck in them and occasionally slipped off narrow rural roads into snow-filled ditches without any damage, except to my ego, and I was stranded in my car overnight in blizzards a few times. I never felt a life and death situation faced me as I was well prepared, although many other travellers didn't fare so well. Blizzard warnings usually drove me to the nearest hotel, one of the better ones if possible, where I holed-up until the storm subsided.

On the road needing gasoline in communities without electricity meant it was sold by gravity-flow from circular, 10-gallon, glass tanks mounted on top of the pump. The gallon markings were on the side of the glass tank that was filled with a side-mounted hand pump before each sale. These museum pieces were standard equipment before electrification, a scant 50 years ago. Purple gas was for farmers' use only, in their trucks, tractors and other motorized equipment, as it was cheaper than regular fuel because governments reduced the sales and excise taxes. Provincial highway traffic inspectors often set up highway roadblocks to check gas tanks for its illegal use. There were times when I had neglected to keep my car's tank filled before launching out on a series of calls with my committee when I had to pull up to a farmer's bulk-storage tank for a few gallons of the tax-free stuff. The inspectors checked my tank many times but, fortunately, I was never caught, having diluted my felony, luckily, with a fill-up of the legal stuff. And speaking of gas, one afternoon after pulling into a service station on the southern outskirts of Davidson and filling my tank I went to the office to pay only to discover two pair of feet facing each other on the floor behind the counter. Obviously the attendant was involved in a high-octane transaction.

On the road all the time I lived in Yorkton, my vehicle was serviced by Merv Campbell's British American Oil Service Station. Aware of the need for it to be in top shape, he and his staff checked it carefully almost every week, including the air-pressure of the spare tires.

On the road could have been lonelier as the government refused to buy AM radios for our cars, considered them a luxury taxpayers could ill afford. However, SPC's vehicles staff eluded the restriction and instead of calling them what they were, requisitioned "low level noise indicators." Government bureaucrats did not understand this nomenclature or, if they did, never stopped the purchases. AM radios helped detect radio interference which may have been caused by faulty power line equipment, a broken or cracked insulator, a damaged transformer or a loose high voltage wire connection, the sources for customer complaints of radio static. Car radios were useful to DOs and linemen for this purpose, but those of us not involved with line work did not need them, except for our enjoyment. I was grateful for mine, useful in stormy weather to keep abreast of road conditions and to help stay awake at the wheel, particularly in the dark hours.

On the road the CBC's radio station CBK was my listening favourite, able to tune it in all over our area. Anyone raised in Saskatchewan in those days can likely still recite the announcers words identifying the station when they said, "This is CBK, with studios in Regina and transmitter at Watrous." *This Country in The Morning* and later, *Morningside*, hosted by Peter Gzowski were my favourite programs by far. Gzowski was on the air when I left home Monday mornings, and every other weekday. His homespun manner, sincerity, genuine concern for and about Canada, his knowledge of the country and desire to make its furthest reaches known to all Canadians made me feel like I knew the man, yet I never saw him in person. My love affair with him continued for as long as he was on-air. While I worked in New Brunswick in the late 1970s and returned to Saskatchewan in 1980, as the Quiet Revolution picked up steam in Quebec, and until after Gzowski retired, I believe he did more to meld this country together than any elected political leader or bureaucrat.

On the road with SPC logos displayed on our panel van side doors was problematic, not that I was ashamed to work for the Corporation, but sometimes it telegraphed our identity when it was not prudent to do so, particularly when we cleared easements and our initial approach to clients had to be tactfully handled. We pleaded our case and after 1955 drove unmarked vehicles. Once when I checked in to the Preeceville Hotel, the owner, unaware of my employer, expounded about an unruly SPC construction crew which had trashed some rooms the week before, wondered aloud whether he would be paid for the damages, plus lost business. The crew, no doubt inebriated, wanted a larger room for their after-work party so, with the hallway fire axes demolished a wall and made one. The next morning after checking out I went to the telephone office and called our property and claims people to quickly settle the matter. They did so with the hotel owner unaware of my involvement.

* * *

ON THE ROAD year-round meant I drove thousands of miles on rural roads, on gravel and ice, in and through snow and mud, over washboard, in water-filled ruts and through or around water-filled pot-holes and frost boils. The latter were a springtime phenomena as the frost and moisture came through the gravelled highway road surface creating a quicksand-like quagmire. When it was impossible to drive around them on the road's edge they simply could not be crossed without paying a nearby farmer for a pull with his tractor or team of horses. There were stories about those mean and unethical types who lived close to these mires who dumped water into them at night to extend their pull-out revenues a few more days. A few times I drove over these frost boils on a corduroy road, logs laid side-by-side at right angles to the travel path. They had been placed there by highway department crews until after spring breakup. Another hazard on unpaved highways was the crossing of gravel ridges left in the centre of the road as the graders moved it back and forth to smooth out the pot holes and washboard. To cross these ridges to pass other vehicles needed care and reduced speed, with loss of control being a real concern, not to mention the possibility of vehicle undercarriage damage.

On the road in late fall, winter or early spring, when most of my annual mileages were driven, was when road conditions were the poorest. Our work regimen of holding organizing meetings, winding down sign-ups and obtaining permission to cross private property were all held after harvest and before spring break-up, when farmers had more spare time for canvassing. So the majority of our miles were in the dark hours of the year. A full moon often helped driving without headlights when the snow was blowing across the road, easier to see without the blinding flash-back glare.

On the road during these cold months when rural roads were clogged with snow I travelled in a variety of modes, bundled up in my survival gear and carrying my ever present brief case. Tractors were the most common, some with but most without enclosed cabs. I also travelled on stone-boats, in sleds or sleighs, some enclosed, drawn by a horse, sometimes two, used every weekday to take the farmer's children to and from school. Had I known how to ride a horse, there were times when this would have been handier than the other options. Bombardiers, assigned to some DOs, were handy – a rough ride but enclosed from the wind and cold when they took me to meet my committees to complete their sign-up work when jobs finished and the rural roads were blocked. Snowmobiles had yet to be invented, but there were a few homemade contraptions, powered by an old car engine driving an airplane propeller mounted at the back – a rather perilous transporter, but they worked, my final dismount from them always with a sense of relief. Their propellers were made of wood and once propelled Tiger Moths and Fairchilds, those early airplanes used to train tens of thousands of Allied pilots and air crews during the Second World War until hostilities ended when they were deserted at the many British Commonwealth Air Training Program airfields that dotted the prairies. Snowplow clubs, with farm tractor mounted rotary blowers were not universal in our area, but where they existed were a blessing, beautiful to follow in the early morning sunshine as they threw their arch of snow off the blocked roads to get me to my destination.

* * *

ON THE ROAD always meant paying attention to the black lettering on the crossed white boards that warned of a "Railway Crossing." Steam-engined trains still connected many communities, some only weekly, but a visual safety check down the tracks both ways before proceeding was essential. I emphatically learned this lesson in 1949 in Tantallon when I travelled for Steele Briggs. A grocery salesman who preceded my visit with a storekeeper, failed to make that safety check at a railway crossing half a mile from town and was killed when the train rammed his car.

* * *

ON THE ROAD, studying maps of communities we were about to electrify made me wonder about the lack of imagination of the surveyors of these prairie oasis that happened as the nineteenth century came to an end. It seemed their creativity was left out on the tracks as they plunked these centres down anywhere from seven to 10 miles apart, to keep the distances manageable for farmers to haul grain and cattle to market at the railway line. Often, more desirous or picturesque locations were disregarded in this rigidly imposed adherence by the railway companies. The land for these town sites was usually provided free by the government, the lots later sold to settlers or promoted to absentee speculators.

On the road, anchoring these unimaginatively laid-out prairie neighbourhoods was the railway station and telegraph office, always on a straight section of track. The station was the nucleus of the surveyed tidy rectangles of lots, blocks, streets, avenues and back lanes of the communities that sprang up in repetitive fashion as the rails moved westward. The thoroughfare which paralleled the railway tracks was predictably called Railway Avenue. Adjoining the station was elbowroom for grain elevators, a stockyard with a chute to load animals in and out of railway cattle cars, a freight shed, maybe a water tower, in larger towns a roundhouse for train steam engines and a "section" house to house its track maintenance workers. The street in front of the station, at right angles to the tracks, the main drag, was called, you guessed it, Main or Centre Street. It was usually a bit wider than all others. Streets on either side of Main were First Street West and First Street East, the second ones in either direction, Second Street West or East, and so on. Where the railway ran north and south, the designation of North and South replaced the West and East. Those which paralleled it on either side of Main Street were First Avenue West and East, Second Avenue West and East, Third and so on.

On the road these small centres of commerce were given the designation of "towns" if their population was 500 or more, "villages" if over 100. The smallest were "hamlets." These rural communities were almost duplicates of one another and most buildings were clapboard: a two-story hotel with a beer parlour, a cafe or two and a laundry, a post office, a drug store, a general merchandise store or two sometimes called variety stores, a bank was often made of brick, presumably to give it a look of security, a combined lumber yard and hardware store, a farm implement and automotive dealer/garage/service station, a coal and wood dealer which later became a bulk fuel and oil agent, a church or two or three and if one was Roman Catholic it was the largest of them all with the tallest steeple, a skating rink with a lean-to two-sheet curling rink, a school with a playground, a couple of grocery stores, a harness and blacksmith shop, a tannery, a feed store maybe as part of a flour mill and a plethora of houses from the more elaborate two-story brick mansions owned by the community's professionals to rows of dilapidated shanty shacks, their foundations and lower walls banked-up with earth to keep out the cold. Nearby, in a coulee was a garbage dump. The towns professionals – doctors, lawyers, dentists and accountants – had store-front operations or were housed upstairs above another commercial premise. From personal experience, growing up in a small town leads me to believe these are places where every kid should grow up.

On the road in our area was the ABC line of communities, the naming of which must have been the source of more than a few giggles amongst the survey crews. The first letter of each settlement was in alphabetical order stretching half-way across the province following the mainline Canadian National Railway as it headed west from Winnipeg to Saskatoon after crossing into Saskatchewan from Manitoba. The pattern started with the first village west of the border, Welby, then came

Yarbo and Zeneta. The first letter alphabet name was then restarted with Atwater, next was Bangor, Cana, (Melville and Birmingham were interruptions and why were D and E omitted?) followed by Fenwood, Goodeve, Hubbard, Ituna, Jasmin, Kelliher, Leross, Lestock (why two L's, and why did they miss M and O?), Touchwood (the second T and out of sequence), Punnichy, Quinton, Raymore, Semans, Tate, Nokomis (out of sequence), Undora, Venn, Watrous, Xena, Young, and finally Zelma. The pattern then restarted with Allan, Bradwell and Clavet before the city of Saskatoon disrupted the cycle. Fascinating! There must have been a different survey crew that laid out Saskatoon's west end where the thoroughfares start with Avenue A, the next is Avenue B, then C and on down through the entire alphabet. Pretty dull stuff as compared to the ABC line of towns.

On the road in larger towns and cities the surveyors seemed more creative and allocated parks, exhibition and recreation grounds. Some street names reflected the origin of early settlers, a distinguished mayor or civic leader, a famous son or daughter, a respected professional or a decorated war veteran. The surveys in some centres optimistically laid out block after block in an area that was just naked prairie. Many were never developed anywhere near to what was envisioned.

On the road in the late 50s saw revolutionary trends in the agricultural community on two fronts. First came larger acreage farms as neighbour bought out neighbour causing the disappearance of many small ones. Secondly came the emergence of factory farms, intensely raising poultry, swine and cattle in confined quarters. These changes came with promises of increased productivity with fewer inputs of labour, but such progress, if one wants to call it that, came at the expense of eroding the thriving rural communities and their economies. This translated into fewer farm families to support these rural centres. So fatal declines commenced in size as well as in influence, starting with a store being boarded-up, then a doctor or dentist left and wasn't replaced, the school closed and buses took the children elsewhere, a drug store or bank left its business for the one in the next town. With each defection the community's importance suffered. Then this dilution became a chain reaction, capped when the railway was abandoned, the station closed, the steel rails salvaged, the elevators demolished and, as they say, the rest is history. A way of life was over. Hundreds of grain elevators, sacred prairie sentinels, have now all but disappeared, succumbed to fewer, modern, austere, huge concrete terminals. With better rural roads and highways many small centres exist only in memory, sadly gone the way of the dodo bird.

* * *

ON THE ROAD with a determined work regimen, heavy workloads and inflexible deadlines meant I found precious little time to stop, talk and listen to the tales of local folklore or enquire into the backgrounds or experiences of our volunteer farm committee members with whom I dealt. Reminiscing about this today I realize my almost total ignorance of their various cultures and contributions to society. With farmer passengers in my car in those days the predominant conversations were always about farm electrification. Undoubtedly they had other colourful stories about picnics and picking berries; making wine and home brew; sleighing accidents of their children on their way to and from school over snowbanks; runaway cattle and horses; blizzards or cloudbursts that raged outside during a weekly dance in the community hall or schoolhouse; fairs and exhibitions of cooking, handicraft, gardening, horse pulls, farm animals; cooperative building bees for barns, churches, schools and houses; violent hailstorms; disasters of fires, floods and farming accidents. Our third of Saskatchewan's farm population in the Yorkton Area had been homesteaded by perhaps a more diverse group of immigrants from other countries than the rest of the province. They came from Finland, Russia, Iceland, Hungary, Denmark, France, Poland, Ireland, Sweden, Yugoslavia, Scotland, Norway and England. Usually these folks left their native countries for a better life in a new land with dreams of freedom, empty spaces and promises of bountiful and cheap farmland. These homesteaders became stewards and guardians of the land that dreams are made on, dreams that were rooted in hope, hard work, tenacity, frustration and love. Often immigrants from one country congregated in the same district where language, habits, rituals and customs, religion and culture could be sustained and shared. I noticed but paid little attention to these features whenever they became apparent – distinctive architecture of their churches, homes and farm outbuildings; communal homes for several families under one roof; wedding celebrations; outdoor beehive-shaped clay baking ovens; steam bath shanties not far from the home's back door. I regret these oversights and on the next page or two are a few colourful possibilities, but there must have been hundreds more that I overlooked.

On the road near Tantallon there must have been next to unbelievable stories about Ezekiel Bronfman and his sons, founders of the famous Seagram business dynasty, who came from that Orthodox Russian Jewish farming community. I knew a bit about their fabled exploits in Yorkton at the Balmoral Hotel, with the liquor wholesale warehouse his son Harry Bronfman built between it and the York Theatre, underneath all of which were the legendary catacombs and tunnels that stored their illegal spirits during prohibition days. Later the family moved on to eastern Canada and to even bigger and much more profitable ventures on the national and international scene.

On the road near Veregin I know there were fascinating stories about Peter Verigin, the Doukhobor leader who came to Yorkton periodically in all his finery, riding in a black carriage drawn by two jet black horses. He built his own hotel, the Blackstone on Betts Avenue, because the others refused his business. Why? I remember seeing their rural, two-storied communal homes that housed several families, built on the square with an enclosed centre courtyard, part of their old country collective farming culture. But I never asked about these observations.

On the road, what about Major General A.G.L. McNaughton, who led the First Canadian Division into battle during the early days of World War II. His family operated the McNaughton General Store in Moosomin for many years.

Although I was a Second World War veteran I took no time to learn more about this distinguished Canadian soldier.

On the road near Cannington Manor, south of Wawota, named after Somerset, an English village, was fascinating as it was established to recreate the folk-ways of the British aristocracy who settled on the bald-headed prairie and wanted to hang on to their earlier ways of life. They built a 22 room mansion, a private race track with imported steeplechase jockeys, a hunt club with authentic foxhounds and thoroughbred horses. They played cricket, rugger and tennis. All of it went by the board when the CPR intentionally by-passed the settlement and built the branch line 10 miles to the south downgrading Manor to a ghost town.

On the road near Brooksby meant an almost obligatory visit with Charlie Vickers, an immigrant Jew. He ran a general store there amidst a settlement of his people. He was a leader for many to get electricity in all directions from his store. His sons became successful businessmen like their father and were community leaders in and around Melfort.

On the road it was usually thought the many districts in our area were settled by European and Asian immigrants. Not so. An Abernethy neighbourhood was settled in 1882, its pioneers coming from Ontario and trekking overland from the end of the steel at Brandon, Manitoba. They homesteaded, paid their 10 dollars for a quarter section, and farmed initially with oxen or horses. These pioneer family names included Foster, Stueck, Gillespie, Barnsley and Motherwell, the latter ancestors of William Richard Motherwell, Saskatchewan's Commissioner of Agriculture in the province's first legislature in 1905. He later served two terms in Ottawa as Minister of Agriculture in Prime Minister W.L. McKenzie King's national Liberal Government.

On the road also near Abernethy was the unusually talented farmer Len Thompson, who designed and manufactured a famous line of high quality, much sought after, fishing lures.

* * *

ON THE ROAD sprouted myriads of tales and jokes of the travelling salesman's exploits with the innocent farmer's daughter. For me, all were myths, as no blue-eyed blonde with a Petty-girl figure ever beckoned me towards the hayloft. I have to admit though there were hotels where amorous chambermaids, desk clerks and restaurant waitresses made themselves available to those in the upstairs rooms. They were not in the role of a paid relationship, but were lonely young women looking for excitement or a ticket out from what otherwise was a rather humdrum life in those small communities. That married men were sometimes involved or that honour, fidelity and propriety was at stake made no difference. Arousal of the girls' desire to get out of town was real and arousal was maybe the way to get out. Knowing the whereabouts of the men upstairs, some alone, some lonely, some on the make, some plastered with too many beers in their belly, were mesmerizing temptations. Then there were motels where, if a particular room's light went on, a phone call from across the back fence or vacant lot came an offer to perhaps a willing philanderer. Pitfalls for being on the road.

Cracking the Tough Nuts

IN OUR WORK we headlonged into two kinds of farmers I choose to describe as tough nuts: those who did not sign-up for electricity and those who spurned the placement of power lines on their land. Fortunately both kinds were in the minority in their farming districts but both were formidable obstacles that needed to be overcome if our program was to succeed.

The best way to crack tough nuts is with nutcrackers, and as some nuts are harder to crack than others different kinds of nutcrackers may be needed. For farm field reps it was agreed the best way to improve our nut cracking capabilities would be salesmanship training. I had learned some of these skills with a reasonable degree of success in the school of hard knocks when I was a salesman for Steele Briggs. Charlie was supportive of the idea and as there were no local training facilities that taught marketing, he suggested a correspondence course from Chicago's LaSalle Extension University. Many of us completed the 10 lessons in 1955. SPC paid for educational training when suggested by management, so our enrollment required only a time commitment. What I learned was helpful throughout my career: knowing how to sell, the buying decisions people make before purchase and how to constructively overcome objections.

* * *

THE FIRST KIND of tough nuts were those who needed convincing that electricity would be beneficial to them and their families. That many farmers did not seize the opportunity for electricity when it was first offered may be difficult, perhaps even unbelievable, for those in today's society to imagine. Certainly this mystified and surprised project committee members when they were unable to garner enough of their neighbours to meet the targeted sign-up numbers. Years when crops were average or better, or livestock prices were higher, success rates improved but, even then, few jobs made it on their own without a final push from me. When the follow-up calls were about to be made I queried the canvassers, wanting to know why their best efforts had been unsuccessful with those neighbours who had good, well maintained buildings or, those with 32-volt wind-charger-battery systems or, those they believed had the money. These tough nut prospects were at the top of the list of hold-outs I would try to convince.

Some non-signers said they didn't want electricity, likely a veiled defence or denial of other priorities, a contemplated move to town to farm from there, for retirement or other work or for easier school access for their children. Maybe they had an onerous bank mortgage or a machinery loan or tuition fees and other expenses for a daughter or son enrolled at university. Perhaps bewildered parents wrestled their disappointment over their only son's greater interest in a career in a city's greener pasture than taking over the two or three generation family farm. Many small acreage, quarter section farmers, simply could not afford the cost, had sell-out thoughts and didn't want to pay for a service from which they would not benefit. These small landowners faced a compounding dilemma of a changing agricultural environment that needed modern, bigger and more expensive machinery. Then there were those who wanted electricity but could not raise the $100 down payment, in spite of full granaries. Then were frustrated further when a friend or neighbour was unable to loan them the cash because they were in the same boat, and if this was the case it was likely the local bank manager would be unresponsive. Finally, it was difficult to not empathize with those who simply wanted to be left alone, did not want changes in their lifestyles or farming practices, folks aptly described by Wallace Stegner in his book *Wolf Willow*, ". . . how surely progress destroys what makes a frontier satisfying." Regrettably, I had to write off all these folks as prospects.

I then zeroed in on the hot prospects. Rather than go-it-alone to meet them I preferred to be accompanied by one or two of my farm sign-up committee members, those who were on friendly terms with our candidates. They helped avoid my tripping over snarly opinions that might be held by these folks on controversial issues like larger school units, Crows Nest grain-freight rates, the grid road system, Canadian Wheat Board quotas or provincial politics. When I was comfortable with my orientation and had a sense of how I would approach these tough nuts, we struck out. On some projects we needed two or three days to make these calls, depending on road conditions, our travel mode and the number of nuts we had to crack.

Our first calls were to those our committees thought were on the verge of signing, important because success with two or three early in our struggle bolstered the committee's spirits, mine too for that matter. We did what my Mother taught me as a child when preparing for school exams: "Do the easy questions first." Some prospects wanted to meet the power man rather than do business with a neighbour, maybe to be seen as the project's saviour. Some had concerns about power poles in their fields, wanted assurance their land would be spared, impossible to guarantee at this stage. Some foolishly let politics stand in their way, wanted no part of this socialist brand of juice, but usually saw the light when told kilowatt hours that pumped and heated their water, preserved their food, powered their workshops and barns, would not surge through their electric meters as red ones when Liberals were in office or green ones when the CCF governed. Some disbelieved power would really come, then conceded and paid the money when it seemed likely. Some were simply suspicious, had survived thus far without that highfalutin stuff, and were not much farther along than James Thurber's grandmother whose thoughts he recalled in his book, *My Life*

And Hard Times, "Electricity was dripping invisibly all over the house from wall outlets." These prospects were the pushovers.

If, after these visits, we were still short of the target number, which was the case more often than not, then our push to crack the tough nuts would be more difficult extractions. My committee members at this point often shuffled their feet uncomfortably as we tried to decide on whom we would call on next, often eliciting the comment, "I'm buggered if I know." These challenges were often not puny obstacles. Some calls were short, others up to a couple of hours. A few were anything but routine, others were stressful, even hazardous.

* * *

THE SECOND SET of tough nuts was those who did not want poles on their cultivated land. This issue was also the most controversial item we had to handle at the organizing meetings. All farmers are sensitive to the placement of obstructions in their fields where they have worked long and hard to clear rocks and brush or filled-in sloughs and pot holes for easier machinery movement and to improve their land. They objected, even when obstacles like power poles were for their benefit, but we had no other alternative but to place our lines in fields, unless some enterprising soul invented "sky-hooks" to hold-up the wire.

SPC adopted the practice to build rural power lines on the east and north sides of the 66 foot wide municipal road allowances, a minimum of 150 feet from the road's centre line. Telephone lines were normally on the south and west sides, and usually already in place along the edge of the road. As the two lines needed separation of at least 200 feet to minimize telephone interference caused by electricity induction from the power lines, our poles had to be on private property in the fields or pastures. On advice from the Department of Agriculture we recommended that shelter-belt hedge-rows of caragana be placed under the lines to help eliminate their inconvenience, and to help trap snow in winter or to deter soil drifting.

SPC expected to receive free permission to build rural lines on private property. If permission was refused and an easement had to be purchased for 13.8 kV farm lines or their 25 kV feeder lines, the amount paid was $4 per pole. Easements were purchased only as a last resort and if the land owner wanted service on that quarter section at a later date, these amounts had to be repaid in addition to the normal AC construction charge. These were local lines to serve their community and if easements were paid universally this expense would have increased the AC cost to those for whom the lines would serve, and this didn't make much sense.

When the sign-up was finished our engineering staff laid out the power line routes. The sign-up committee was then sent easement forms along with maps of the job and was asked to have all the affected landowners sign the papers to give SPC permission to cross their property without compensation. This free easement was sought on all quarter sections where poles and anchors would be placed except the home-quarter section of new customers, as they automatically gave this permission as a condition of taking service. About 15 per cent of the forms on most projects were returned unsigned, with permission-to-cross refused for a host of reasons, including absentee owners who lived in other faraway communities or were living in another province or out of the country. These were mailed written explanations of our intent and were handled by our office.

The non-signers who lived within or near the project were more difficult to cope with than the hold-outs we encountered in the initial sign-up phase. My role was to find out how serious their objections really were, to sort of run interference for our construction forces. The crews needed to know beforehand who had agreed to have poles in their fields and, on the other side of that coin, where they might encounter landowners that could be violent or might otherwise delay their work. My visits with these owners to convince them of our plans often boiled over into outrage. They felt we were badgering them after the committee had already applied local pressure.

While this work was not dear to my heart I rarely lost my cool. When tempers flared I just listened passively and allowed them to vent their spleens without any interruption, making mental notes of their points of argument. Then, with their spring unwound I reentered the confab. I likened it to repairing a hot boiler under pressure: until the steam has been vented and it has cooled off, repairs cannot be started. When facing these awkward confrontations my feelings have best been described by the beloved prairie author, W.O. Mitchell, who wrote, ". . . It's enough to give a gopher the heartburn."

Most farmers accepted the lines as a part of progress and the evolution towards a better way of life, albeit with some reluctance. Sometimes, those getting service refused permission on other land they owned particularly if they farmed large acreages and thought they had more than a fair share of poles and anchors. We tried to accommodate some with line reroutes, but this meant others would be affected. Usually these changes were as unpopular and stirred-up the pot as much as the original line proposal. Those who would get service were not as difficult to convince if the routes could stand the test of logic and common horse-sense. The hard cases were those we would not serve, absentee owners who farmed from town or rented their land and already enjoyed the benefits of electricity.

An Aylsham district farmer owned some 55 quarter sections of land, more than 13 sections, and an array of field machinery to handle this huge acreage. A local story one year after harvest was about an acquaintance who met the large landowner and asked when he was going to harvest the crop still standing on such and such quarter. He apparently had overlooked it and after lunch several self-propelled combines appeared in the field and, like watching an elephant ballet, they moved ponderously and methodically around the field and by nightfall the crop was under cover. For three years I was involved in projects in this district. It seemed that with so much land it was inevitable his property was affected. He always refused a committee-mans overtures for permission-to-cross. This meant I had to see him. His refusal was not a serious objection. He simply did not want the lines on his land but on the other hand did not want to stand in the way of others getting a service that he already enjoyed on his home farm. I never got his signature and our crews never encountered a problem with him while working on his property.

On the positive side we frequently had suggestions from the committees after they had been lobbied by their neighbours to move a line a bit this way or that, or to relocate it to a pasture or in poorer, low-lying land or to follow an old creek bed. These changes were readily made. There were instances where existing power lines running through pastures acted like lightning rods and protected cattle from death and injury during electrical storms. Albeit these cases were not widespread I always mentioned this benefit when it was appropriate to help my case.

A tactic I sometimes used with a hold-out on an easement issue when all arguments were exhausted and we seemed to be at an impasse: I'd say, "Let's change roles. I'll be the farmer. You be the power man. Where would you build this power line?" If the route was reasonably located and the inconvenience equitably shared amongst his neighbours, this often jarred them into reluctant agreement with, "Oh, okay, where do I sign?"

* * *

WHAT FOLLOWS are some of my adventures to corner and crack some of the tough nuts on my projects so their districts could receive power and with the fewest objections from neighbours.

* * *

THE "HAMPTON-DONWELL" project failed in its 1952 attempt to get electricity for lack of sufficient sign-ups. This defeat bewildered us as theirs was a good, mixed farming district. So, in 1955 we tried the job again.

The day I spent in cleanup work revealed the "why" of their earlier collapse. I had made the last three calls by myself to prospects the committee was uncomfortable about, two of whom signed. After supper I met my farm committee in Hampton's town hall. Our quota had been met. When told this good news their reticence vanished, the bootleg booze emerged, so fast it made your head spin "The kind without a government label," they said. Then they fessed up.

This district was a Liberal stronghold. All my committee members were card-carrying Liberal members and passionately deplored the CCF. They had disbelieved me when told at the organizational meeting "Their politics didn't matter. My job is to help you get electricity." Their suspicions three years earlier lead them to believe they could not possibly get power, desired so badly, from a government they consistently voted against, so they had just gone through the motions, then complained. With that out of the way, we partied.

Earlier that cold day we called on what I thought was a hot prospect by the name of Krepakavitch. He reportedly had the money, owned a 32-volt lighting plant, fed and watered a large cattle herd the back-breaking way and electricity was high on his young family's list of priorities, but he had firmly withstood these influences. Two of my committee came along and as the road to the farmyard was blocked we walked about 400 yards over the snowbanks to the house.

When inside I pulled off my toque, unzipped my parka, and when introduced as the power man Krepakavitch instantly flushed with anger. He yelled at my companions, "I told ya I didn't vant power from dat bunch. I sad don't bodder me. Get out!" His hell-bent action was toward a corner closet, a clear signal for my allies to scram. They bellowed at me, "C'mon Dave, let's get outta here. Fast!" knowing the closet stored his guns. They grabbed at me on their way out the door so I bolted too. Then, in disbelief, I saw his rifle.

I rounded the corner of the house on the dead run, parka flapping, struggling to put on my toque with one hand, briefcase clutched in the other, my companions well ahead racing toward the car. I glanced back. There he was. Gun to his shoulder. Aimed at me. Two shots rang out. The bullets zinged past.

Intentionally over my head as a warning, or had he aimed and missed? With the crazed, armed man at my back the run to the road may not have been an Olympic record but I would likely have made the team. In the car we breathed a collective sigh of relief. I was more than perturbed at not being cautioned, but my cohorts staunchly believed he would not be so irrational.

This brown-trouser adventure scared the begeebers out of me. It ranked right up there with my first taste of "Action Stations" on the North Atlantic in the Second World War. I had an acute appreciation of how Winston Churchill felt when he wrote after his South African Boer War experiences, "Nothing in life is so exhilarating as to be shot at without result."

At my committee's town hall party we joked over the incident. It never crossed my mind to report it to the RCMP. I simply accepted it as the adventurous part of the job.

The fanatical Krepakavitch relented, applied for and paid the full AC price and got electricity a year later.

* * *

IT WAS A WINDLESS, hot, July day in 1956 when I toiled on the easement hold-outs on the "Hampton-Donwell" job. I was accompanied by a committee member, one who was with me on the Krepakavitch visit. Our first call would be to a quarter section farmer who had not signed up. He reportedly had no money or a desire for service. He had also told the committee he did not want any poles on his land.

Something was suspicious about the trail that meandered through a couple of hundred yards of bush towards the tattered buildings, after we had gone through the barbed wire gate. It was not a well-travelled way but I made no comment to my colleague. The only vehicle tracks were lug imprints from an old-time steel-wheeled tractor.

As we entered the farmyard an assortment of dogs announced our arrival as my car shooed a flock of chickens out of the way. We stopped near a tiny, unpainted, clapboard, two-room shack. The dogs did their thing on my car's tires and several scantily clad, or entirely naked, and filthy children stood in the dust and dirt and stared at us curiously in a bewildered sort of way. The farmer's wife appeared in the open doorway, a robust lady, barefoot and also in need of some soap and water. She was dressed in shorts and sported a faded white T-shirt, stencilled with the front cover mastheads of several current woman's magazines: *Ladies Home Journal, Women's Home Companion*, and *Good Housekeeping*. Two holes extruded her bobbling bare breasts, large and unencumbered.

After a nonchalant exchange of "Hi"s, certainly staged on my part, I stuttered out my desire to talk with her husband, trying like the dickens to be unfazed by the two 38s aimed right at me. We got out of the car and she ushered us through the kitchen to their home's second room, a combined living and communal bedroom. With no door or window screens the children, flies, mosquitoes, dogs, cats and chickens all had unrestricted access to the rooms and their meagre belongings. We sat on rickety wooden chairs, my companion near a curtain-less window from where he could see the yard. My back faced a wall opposite the door through which we had entered. She left to call her husband from a nearby field from where we could hear the pumph-pumph-pumph of an old John Deere.

Shortly the tractor's engine noise grew louder, then stopped. The clomp of heavy footsteps prompted my colleague to say, "Oh, oh, here comes Adam." Before I could question him, as this was not the man's name on the easement form, the clomps entered the kitchen and made their way towards the door. There, full-frame in the doorway stood a bare-headed, giant hulk of a man, well over six feet, tanned as black as Toby's arse and stark naked except for his clodhoppers.

He had been harrowing a summer fallow field, a dusty and dirty job on any hot, dry day. His entire body was covered in thick, black grime, glued-on with perspiration. This bizarre human form was accented by a pink mouth, the whites of his eyes and etched, muddy, sweat rivulets that made their way down his brow, neck and torso. He emitted a knowing grunt at my colleague, grabbed another chair, placed it about four feet in front of mine, then plunked himself down directly facing me, his legs splayed spread-eagle-fashion with a hand on each knee so that he brazenly displayed his well-endowed, grimy genitals. His eyes peered eerily into mine as he blurted out, "Whaddayawant?" Had I been ambushed?

Slack-jawed at the sight, somewhere between terrified and thunderstruck, I could not believe the apparition that confronted me. I took a deep breath, tried to collect my thoughts. When told, "I want to talk to you about the power line we will build on your land this summer to serve your neighbors," he became upset, anger visible on his filthy face. Transfixed to my chair I glanced furtively at the open window. Could I dive through it and into the safety of my car before the galoot could get at me, briefcase be dammed? My companion intervened and the brute calmed down. He said, "It's not okay for de lines on my land." I thanked him for coming in from the field, made no attempt to coerce him further and, without any hesitation, slithered outside like a startled snake finding refuge in a rock pile.

When away from the buildings I admonished my associate, obviously not weighed down with intellect, that I would have liked to have at least known the farmer was apparently harmless when push came to shove. My cohort doubted I would have gone if he had told me what we might encounter. He was right, it was a believe-it-or-not episode I had not needed. While it was a laughable experience it was not funny at the time.

This couple was a pitiful, sad case. Both adults were mentally disadvantaged, eking out a living and bearing numerous hapless children, but in spite of my personal standards they both seemed to be quite comfortable in their skins. My advice to our construction crew was to be cautious when building the line on this quarter section, that I believed the owner was not a safety threat. I later learned the line went across his land without incident to the self-help crew.

* * *

IN THE DEAD OF the 1956 winter, the "Nut Mountain South" project of about 30 farms was three short of its target. Two committee men and I started our rounds early one morning on a cab-less tractor, our mode of transportation over the snow-clogged roads, one of us perched on each fender. We were able to sign-up the first prospect. My colleagues were encouraged as we left to make more calls.

As the day wore on our success rate diminished. In fact it was a failure, and we were running out of potential clients, down to the last two. Both were essential, but they presented a formidable task. One, a couple in their mid 60s, reportedly had the funds but intended to sell their half section farm and retire. With modest, well-kept buildings and good field equipment, the prospect of an easier life in town was appealing. The other, just across the road from the first, was a young couple with two infant children who had just moved back to their vacant farm from town. Their belongings included several, now useless, major electrical appliances they had used in town: washer, dryer, fridge and electric range. They desperately wanted electricity but did not have the $100 down payment, in spite of overflowing granaries, unsalable because of restrictive grain quotas, plugged elevators, a constipated railway transportation system and overflowing seaport loading terminals.

My sales pitch to the elderly couple of needing two more signers elicited the commitment, "Well, if the line goes past our farm to serve the new folks across the way we'll sign for electricity too. It will likely increase the value of our farm if we sell it and, besides, we don't want to stand in the way of our neighbours who are so anxious to get it." This last comment played heavily on their minds, not unusual for community minded people, a statement I had heard many times and, admittedly, had used to extract other sign-ups. We left with their promise ringing loudly in our ears.

As we climbed up on the tractor I asked, "Can either one of you loan this prospect the $100 with the stored grain as security." While they were willing neither had the money. We were so close to success. I racked my brain for a solution, feeling

I had head-longed into a wall that couldn't be climbed. On our way across the road to the other farmyard, showing its years of disuse, house and barn faded to a weathered wasp-nest gray, a brain wave entered my head. I remembered Ben McChesney, a well respected registered seed grower who farmed south of Kelvington, just a few miles to the west. I had known him when I worked for Steele Briggs and remembered him as a decent man with a lot of grain storage capacity in his farmyard, needed for his seed cleaning business.

When in the young couple's kitchen we discussed our respective situations, their burning desire and our desperate need. There was no doubt they wanted to get back to a life with electric lights and, at the same time, help the project and their district. But, from where would the money come? I wondered aloud if McChesney, when approached, might help them in some way. They readily agreed to my proposition so I phoned Ben. He remembered me. I explained our predicament. He said, "If some grain could be delivered, my bins have room and I'd be more than willing to help them out and give them the $100." The young couple were elated. Now to deliver the grain? The farmer called the RM secretary, explained our puckersnatch and requested their snowplow clear the municipal road and a pathway to a granary. He agreed to help. The sun's shadows were long when a Nut Mountain trucker followed the plow to the bin, loaded and took the grain and the farmer to McChesney's. The farmer's wife made us lunch while her husband was away.

The deal worked. By early evening he signed up as soon as he got back. In the dark we returned across the way to the elderly couple's home. I showed them their neighbours cheque, receipt and signed contract. We admitted to our afternoon capers of having the road cleared and the grain hauled. I sheepishly said we desperately needed to hold them to their earlier commitment. They signed the forms without hesitation and paid the construction charge in full. We thanked them and, both pleased and relieved with our success, climbed back up on the tractor and, in the dark, headed for Nut Mountain where I had left my car. That ride didn't seem as cold as was the temperature.

Frequently over the following months I had guilty feelings over this deal that I had pulled-off with the elderly couple but in wrestling to clear my conscience I never came close to winning the match. The next September en route home to Yorkton one afternoon, it got the best of me and I pulled off #49 Highway and went south on the country road to their farm. The power lines had been built a couple of months earlier, the fresh earth piled around each jet-black, creosoted pole and the shiny porcelain insulators were a dead giveaway of the line's newness. I could see the wires fanning-out to their buildings from their yard pole and through the kitchen window gleamed a new refrigerator and electric range. They did not recognize me without my parka, boots and toque. I reminded them, "I'm the guy who tricked you into taking power last winter." With broad grins they beckoned me in and he told me about the last night of harvest a few weeks earlier. He said, "The combine broke down with only a couple of hours of work left. I brought the machine into the yard near the transformer pole and under the light made the repairs, returned to the field and finished with all the crop in the granary before rain started in the morning." She proudly showed me the new equipment gracing her kitchen plus a washing machine, toaster, other small appliances and the bright lights. They invited me for lunch, a pleasant time chatting about our respective families. With prospects of an easier life with electricity their retirement plans had been put on hold. They were very thankful for my actions the previous winter. My conscience was clear.

* * *

EARLIER I MENTIONED staying two nights during the winter of 1955-56 at the farm home of Francis Eyre, the committee chairman of the "Fosston West" job. My help was needed to complete their sign-up. We made many calls with a reasonable degree of success and towards nightfall on the second day we needed one more. A prospect on our list was an old-time farmer by the name of Christianson. It was generally believed by the committee that he had the money and, as well, he owned a fine set of buildings served by a 32-volt wind-charger system. Every canvasser had been met with so many questions and objections the committee had given up until my arrival. We needed him badly. Francis called and arranged for us to see him that evening.

After dinner and in the cold on that cloudless, moonlit night, bundled up with fur robes over our winter gear, we bobbled along in silent misery on the wobbly one-horse stone-boat up and down over endless snowbanks that plugged the roads for about six miles to the Christianson farm. We followed the road allowances, a longer route than taking possible short-cuts through fields, Francis was concerned about tiring the horse in the event it broke through the deep, crusted snow and having to plod through it or running into buried barbwire fences that might injure the animal. Our stance on the stone-boat tested my sense of balance which, from my Navy days, I thought was pretty good. But I had to hold on to him for support, as he was more stable because of the reins. The horse's hooves squeaked high-pitched, eerie sounds as they dug into the frozen snow with each step, the only noise under the bright moon and twinkling stars in the frozen countryside.

One of the Christianson sons met us at their big red barn and helped unhitch and stable the horse. He led us to the house, a two and a half storied, well built brick home. In the back porch change room, permeated with the distinctive odours of farm chores, we unloaded our fur robes, parkas, toques, mitts, scarves and boots which joined their family's array of outer garments. In the spacious, warm kitchen we met the family patriarch, his wife, the other two sons and their three wives. Our hands thawed out, wrapped around hot coffee mugs before they were drained to warm our insides. I was invited to a seat with my back to the wall at a large, rectangular table that sat at least a dozen people. The father faced me directly opposite, Francis was at the table's end to my right. Across the room, clearly in my view, the other seven adults took perches on a counter of a long, built-in cupboard, their feet dangling idly above the floor. Beside them under towel-covered plates was a prepared lunch. In another room off to my left with a curtain in place of a door, guttural sounds emanated, I presumed from a handicapped person.

Christianson was about 65, a large, stocky, ruddy complexioned, grey haired man of few words. He said, "I'm not at all sure about taking electricity. I have a lot of doubts and questions." He was told I would do my best to answer so, one at a time, they were launched. I answered, slowly and carefully, presented the information as convincingly as I could to overcome his concerns. His queries were as common as mosquitoes beside a slough in summer as I had heard versions of them all and they had been explained at the organizational meeting. Each time I finished and showed I was ready for the next one the seven sets of legs across the way swung back and forth as he pondered my answer, then stopped abruptly when he posed the next question. This seven-member pantomime repeated its performance frequently over the almost two hour interrogation. There had not been one word, a query or audible sound from my counter audience the whole time.

The group of seven wanted electricity badly but the unquestionably in-charge father needed convincing. Finally, his thoroughly wound-up spring ran down and, with his objections satisfied, he said, "Okay, I guess I'll take it." I could not help but hear Francis's sigh of relief, our hard-sought-after target had been reached. The folks across the way tried to contain their glee as they leapt from the counter and prepared to serve lunch. Their bridled joy no doubt would burst forth the first time they were away from their grand "old man."

I completed his application form. He signed the contract. He said he would pay cash so I started to make out the receipt for the full payment as he went to the curtained bedroom and returned with a Chase and Sanborne coffee can. He took off the lid and there, exposed in full view for all to see was a rolled wad of paper currency and cheques. The AC construction charge was an uneven number, like $472. He rummaged through the grain cheques and produce vouchers trying to find the right combination of bills. As they were mostly 20s interspersed with 50s and the odd 100 he was unsuccessful. Neither Francis nor I could provide the correct change. I suggested a cheque payable to SPC. This was not possible as he did not have a bank or credit union account, didn't trust or believe in those institutions. He twice went to the bedroom and returned with more cans, each jam-packed like the first. Finally, the right combination of bills was found.

I don't know how many containers there were on the shelves in that back room or how much they contained but I thought at the time upwards of $5,000, maybe more, in each. With such strict control over his family and as well, from the nature of his questions and actions, he was suspicious of everyone and yet I found it odd the location of those containers was revealed to me, a complete stranger. He must have trusted both Francis and me. The reason for his distrust of banks, and the even more friendly credit unions, could possibly have arisen from some loss he had suffered, maybe a locally owned bank failure before Canada's Bank Act came into being 25 years earlier that brought an end to these unstable, small institutions and shady money lenders. Maybe it originated with dishonest grain or livestock buyers, an implement dealer. Who knows?

After finishing lunch we bundled up in our gear, hitched the horse to the stone-boat and with the moon lighting our path we wearily made our way home in the cold, arriving well after midnight. Next morning after breakfast Francis delivered me to the Hendon railway station to catch the train to Wadena to retrieve my car.

Francis Eyre later became an executive member of the Saskatchewan Association Of Rural Municipalities, a powerful rural organization. I met him several years later after an address I had given at one of their annual conventions in Regina's Trianon Ballroom. We reminisced about our Christianson adventure on that cold night.

* * *

THE CHRISTIANSON CASE at Fosston was not the only prospective client I met with confidentiality concerns over their economic status, perhaps wanting to evade taxes or simply were mistrustful of banks, credit unions or bookkeepers having knowledge of their private financial affairs. Most were usually discreet on the whereabouts of their tobacco tins, coffee cans or jam jars. Always they contained negotiable produce cheques or vouchers, some several years old, for grain, eggs, cream, milk, poultry, cattle or swine. When the construction charges were paid with these wrinkled and plain looking pieces of paper we were leery of their validity until they had gone through the banking system clearing houses, but I cannot recall any that bounced.

When these sorts of farmers agreed to sign-up, the final bit of business was their payment and my issuing of a receipt. He would then disappear to the barn, maybe its hayloft, the machine shop, chicken coop or granary to retrieve his safety deposit box: a Nabob coffee can, a Squirrel peanut butter tin, a Keen's mustard jar. One farmer's hideaway was the barn's manure pile. He was not able to untwist the jar's lid so brought it into the house. The odours it emanated told me its repository was no mystery. The metal lid had rusted so badly it had to be pried open with pliers to retrieve its contents, accompanied by the owner's sheepish grin. I had experiences with more than one case like this one.

* * *

GEORGE LINCOLN'S "Wawota Rurals" project committee ingeniously helped prospects who were cash-poor. To overcome this stumbling block early in their canvassing work the committee members put up what they each could afford into a trust fund. It amounted to over $5,000 from which they gave interest free loans. As their sign-up target of new farm applications was exceeded there was no need for me to attempt any extractions of non-signers. They had very capably done their work. As a matter of fact, to have even suggested I should call on some who had not signed would have been an insult to their thoroughness. Lincoln confided the trust fund's existence only after the sign-up was complete and Head Office had sent out the wiring notices giving assurances to all signers the power lines would be built the following year. Two years later he took great pleasure in telling me that every cent had been repaid, so typical of the innovativeness, trust and cooperative nature of prairie farmers.

Doubtlessly, and usually unbeknownst to us, similar financial assistance was given in some way by hundreds of committee members or other neighbours. These kind, considerate and generous acts were normal treatment in those days from one farmer to another.

* * *

I WAS WORKING ALONE clearing easements on the "Carrot River East" project and had entered a stubble field where its owner was summer-fallowing with a one-way discer. The nearby farm buildings were vacant as the farmer lived in town, already enjoying the use of electricity. These were benefits his rural neighbours now wanted, but the committee members had been flatly rebuffed when they had asked for permission to build a half a mile of line on his land. I parked my vehicle near to where he would arrive as he completed his round in the field, got out of my car and waited for his rig to approach. The machinery stopped but he did not turn off the tractor's engine which seemed a bit strange. I approached the tractor but he made no move to dismount, just sat there with an unlit roll-your-own hanging from the corner of his mouth. He turned around and looked at me without any expression, also an unusual action or lack of it. I climbed up on the tractor tow bar and supported my stance with one hand on the wheel fender, the other clutching my briefcase.

After introducing myself he responded with a cool grunt. Then I said, "I'm from Sask Power and I want to talk about the power line we need to build on your land." Without saying a word or any warning he jammed the tractor into gear, pushed open the throttle and released the clutch. The machinery jerked forward. I lost my balance and, as I started to fall backwards I jumped, fortunately landing on the short side of the one-way's hitch and fell, unhurt, to the ground. I had escaped the ominous discs as they sped past but this uncivil dolt u-turned and came after me. Scared and all a-tingle I got up and ran lickety-split to my car, not waiting for any oracle from Delphi to tell me to get out of there. My car's engine started without a miss and I va-roomed across the field, through the open gate and to safety. I went down the road a-piece, stopped, collected myself, calmed my thumping heart, pondered the close call as I recorded my comments of the incident for the benefit of the construction crew. Then I headed out to deal with whatever the rest of the day cared to deliver, hopefully obstacles that would be less combative.

These were the kind of ornery critters, a small minority, to whom our legal beagles served registered letters of intent and, if needed, expropriated their land for as long as was necessary to build our lines. These preliminary contacts alerted the construction crews where and when to be cautious about their personal safety.

* * *

ONE AFTERNOON in 1957, Ted Durnin, engineer in charge of construction, unexpectedly called me. There was an emergency a few miles south-west of Yorkton where a new 138 kV high-voltage transmission line was being built from Saskatoon to interconnect with the new Red Jacket line substation south of the city. An armed landowner had stopped the pole-setting crew in its tracks as they were about to enter his field. Durnin asked, "Dave, could you go out and resolve whatever is the problem so work can resume?" This was not part of a farm field reps job as property and claims people normally handled these matters. However, they could not get there that day and the costs of an idle high-line crew were substantial. I was concerned and uncertain, but Durnin expressed his confidence in me, so I reluctantly agreed to try. I called my RCMP neighbour friend, Ewen Gray, and requested he or another officer accompany me. He declined as they could not become involved where a property owner was being invaded against his will, unless and until he fired his gun. An armed defensive stance did not constitute sufficient grounds for police action. I was disappointed and became more apprehensive. In fact, I was scared as I left the city.

The crew vehicles were parked on the rural road at the point of entry into the field. I talked with the foreman about the farmer's actions and the crew's plans. Then I entered the farm yard, got past the dogs to the house and hesitantly knocked on the door, covered in goose-bumps and fully expecting to meet an angry, perhaps even armed, man. Even though my car was unmarked he must have known who I represented. The door opened, no gun, no anger. We shook hands. He invited me into his kitchen where we sat and talked. He was reasonable and pleasant and, while upset, he was calm. He knew of our plans, had received the registered letter and understood we had the legal right to be on his land whether he liked it or not. What stuck in his craw was no one from SPC had taken the time to come to see him when he had questioned our intentions and wanted to discuss his point of view. He felt he should have had that courtesy but knew any suggestions he might now have had was too late as the work was too far underway. I agreed that the Corporation had run rough-shod over him. I asked that he let our crew continue. If he would, I promised he would be visited within a day or two, the easement and compensation rates for this kind of transmission line would be explained and paid. He agreed. I asked to use his telephone and called Durnin. He promised the property people would do as I had indicated. The farmer seemed satisfied.

I left, then asked the crew foreman to use extreme care and respect for this property while on this land. I returned home, relieved. This incident and later my self-help scheme involvement with Durnin established a special relationship between us which I valued until his 1968 retirement.

* * *

WE PLANNED to electrify about 100 farms on the "Mikado-Veregin" project. An important part of this job was a 20-mile 25 kV line that would extend east of Canora to Kamsack, plus the connection of citizens in the two villages of Mikado and Veregin. This line would interconnect with a similar one that ran north of Wroxton and the two would serve the town of Kamsack, one backing up the other as a more reliable source of supply in times of trouble.

This project was in the heartland of the Doukhobor community. These were settlers who had emigrated there in 1899 to escape Russian persecution. One group disgorged from a railway train at Yorkton in mid-winter had resolutely trekked on foot more than 30 miles, women, men and children with their belongings on their backs. These settlers and their second generation offspring were a determined and hard working people.

Organizing this project had been difficult as many families still spoke only their native language. Organizational meeting proceedings and subsequent discussions with many of the farmers needed translation. On top of this we discovered midway through the sign-up that several of our committee members were Communist party members, which did not endear them to others in the district who believed differently. Many days were spent on this project in our hard-sell mode trying to crack the tough nuts and obtain the required sign-ups yet, in spite of this, the number who had paid up was far short of the target.

SPC operations people wanted the 25 kV line to Kamsack for system stability and reliability reasons, yet company management didn't want it built without farm support along the way because without them the investment per signed-up customer was almost double the provincial average. Sarsfield, when made aware of the stalemate, insisted an appliance demonstration, aimed at the distaff side of the farm community, would assist the stalled sign-up canvass. He was certain if the farm wives were convinced about electricity's benefits their influence would carry the day. It may have been a nice theory for some parts of the province but Harry Jessop and I strenuously argued against the idea as we knew the area. While these farmers had the latest field implements and machinery, modern barns, workshops and tools, some of their houses were still sod buildings with two-foot-thick earthen walls or log homes chinked with mud and dirt floors, buildings which local electricians had difficulty wiring to meet the provincial safety codes. There was no doubt who wore the pants in those families. Pressuring the farm wives was not a solution. The women still used outdoor, bee-hive shaped, clay baking ovens to cook their bread and pastries. They still used wash tubs and scrubbing boards for their washing chores. Many homes had exterior steam-bath shacks, part of their rural peasant culture where, in winter, they heated themselves to the bone then youthfully cavorted in nearby snowbanks.

Sarsfield persisted and, with reluctance, Jessop and I coerced Kamsack and Canora electrical appliance dealers to fill the Veregin Town Hall's stage with ranges, refrigerators, washers, dryers, freezers and small appliances. On the appointed afternoon the local committee rounded up as many women as they could, perhaps about 50, most adorned with their lovely colourful babushkas. Lillian Vigrass, Penny Powers, SPC's home economist, accompanied Sarsfield. He was dressed in a dark suit and tie and in his usual flamboyant style opened the session with gusto, preaching along the theme, "Here's what electricity will do for you! When you go home be sure you get your husbands to sign up!" Penny Powers did her best to instill practical homemaker ideas and interest, but the two of them could not be understood by many of their audience, having refused our suggestion they use a translator. The women patiently and politely sat throughout the affair with stolid looks on their faces. For some, hot and cold running water and an automatic washing machine was quite a long way from the bucket and scrub board they still used, or a pail suspended in a well as a refrigerator alternative. Jessop and I sat at the back with our committee. Many of the ladies did even not bother to go on stage afterwards to look at the appliances. The afternoon was a bust. Afterwards, Sarsfield conceded we were right.

Our sign-up efforts continued without much success. Finally, the size of the project was reduced by eliminating territory on the north and east where the sign-up had been the weakest. That part of the job from Canora to Kamsack was salvaged so the 25 kV line could proceed. In the area excluded were several farmers who had signed up so there was no other alternative but to refund their money and they would have to wait for another opportunity to get electricity. The final job amounted to about 70 farms and the two villages.

The suspicious nature of these Ukrainian people towards electricity and the farm sign-up committee resulted in an unusually high percentage of "no permissions" to build our lines, a number we were later unable to appreciably reduce.

When construction started on this project the following summer we were called by the crew foreman one Friday afternoon. The pole crew had been stopped by a farmer who had not signed for service and had refused permission to build a mile of line on his land. The manner in which he carried out his objection was frightful. He had lined up his wife and several young children against his barn wall and threatened to shoot them one at a time, like clay pigeons in a shooting gallery, if our workers came through his field's fence. The foreman asked we come immediately, to which Jessop and I responded and got there within the hour.

By the time we arrived at this unbelievable, bizarre scene others too had gathered. The farmer's wife sobbed, wailing loudly as she rocked back and forth. The children's whimpers were almost tearless the ordeal had gone on so long. The youngest toddlers clutched at their mother's skirt while several women chanted and talked quietly in Ukrainian, some distance away. Three or four men went as close to the possessed gunman as they dared, to talk and try to subdue him. He stood alone, a little ways off, his eyes glazed, grim intent on his face, a double barrel shotgun cradled in his arm at the ready, a coat pocket full of shells. It seemed there was little doubt he would carry out his threat if our crew moved. The police would not intervene as we were the intruders on his property. This was a tough nut situation if there ever was one, much like a ruthless war zone incident.

Jessop and I talked with those members of our farm committee who were there and came to the conclusion the best solution was to assemble a crowd of the farmer's close friends and neighbours about to receive service and divert his wild intent. Word of the potential disaster trickled out as the committee drove to spread the word and recruit by telephone those folks who had these instruments in their homes. Cars started to arrive with a goodly number within an hour. As well as those wanting to help there were the inane curious, that element of society who cannot resist to gawk, as if it were an entertainment extravaganza, the grave misfortunes befalling their fellow man.

The crazed farmer's male friends and neighbours gathered near him, speaking persuasively in Ukrainian. As if on signal they moved in and surrounded him with their protective wall of bodies. The women seeing this, moved to encircle the family at the barn wall. The men disarmed and led him into the barn. He seemed relieved to be set free from the stress and mental anguish that had gotten away from him. The entourage of women and children headed quickly for the house.

Reports from the barn indicated a defused and calm situation so with our committee's concurrence we decided to take advantage of it and have the construction crew move in. They quickly dug the 14 holes and set and tamped the poles, the sounds of the digging machine clearly audible in the farmyard. I suspect there was never a faster mile of rural power line poles erected and the wire strung anywhere in the province, before or after, than happened that evening. The crew finished and headed for their campsite. The crowd dispersed and as darkness fell conditions resumed to normal. We left too with the assurance some of the committee would stay until they were sure the farmer and his family was safe. There were no further incidents.

This farmer was converted into an enlightened believer a year later when he requested and received service, having to pay $50 more than his neighbours.

* * *

IT WOULD BE wrong to leave the impression that I, and the other farm field reps, daily faced dangerous or life-threatening encounters with radical individuals who lost their cool, or that rural Saskatchewan was overridden with eccentric and fanatical tough nut farmers. The stories related here were the extreme cases but they did happen and are etched in my memory more deeply than the many thousands of others who cooperated and warmly welcomed our efforts.

I tried hard to stave-off ugly collisions but many farmers became angry with me, yelled obscenities, called me unprintable names and things and, on occasion, scared me to the point I didn't know whether to, as they say in Saskatchewan, "blink, shit or go blind." Some wanted to chew me up like a mouthful of Kibbles-n'-Bits and spit me out on their manure pile. Others, with more guts than a slaughterhouse relished the opportunity to degrade me and the CCF, who were the very ones helping them get electricity, a stance there was no point to defend in spite of personal sympathies at the time to the contrary. Yes, there were some close personal injury situations but no one ever clenched a threatening fist to throw a punch at my valued dental work. I know I was as popular as a skunk at a garden party when some walked away from me when prevailed upon to discuss an aspect of farm electrification on which they were in disagreement. Many who did not sign-up for service or refused permission for power lines on their land were calm, polite and decent people. We simply agreed to disagree. With the thousands of individuals with whom I interacted it was inevitable some would dislike what I was trying to accomplish, but they were, thankfully, in the minority. I regret upsetting those I did and apologize for having caused them any anguish.

On the other hand, I became friends with many of my committee members and I have no regrets whatsoever about those families I coerced into signing-up for electricity. For some, I admittedly put-the-heat-on-high and brow-beat them into submission but I believe none of them would now regret having acceded to my pressure-power-plays. I knew there were just too many advantages in having electricity on their farm.

Lights Up Wawota

NO DOUBT every farm family privately celebrated in some way when, for the first time, they turned on an electric range element, a kitchen light, a radio or outdoor yard light, made ice-cubes in a refrigerator or watched an automatic washing machine do its chore. Similarly, many rural neighbourhoods and communities recognized these milestone events with more formal festivities. Some had a civic banquet, held a potluck dinner, a street or barn dance, a school concert, a community hall celebration or similar event. We were usually unaware of these festivities but later heard about some of them. Doubtlessly, many of our project committee members silently celebrated when the power lines were energized and they could take pride in the significant role they played to bring electricity to their neighbours and themselves.

Some of these celebrations were initiated by our construction crews. Mid-week, with planted poles still several miles away, an enterprising foreman would drop a hint that a dance might be held in town on Saturday night to celebrate, with proceeds to go to a local charity. Maybe he also hinted the town hall could be wired for lights and to power an orchestra's instruments. It could be assumed most of the construction crew, 15 or 20 men, would show up for such a ready made party. Somehow the lines were completed, enough at least to enable the event to proceed. One foreman told me after they had sowed the seeds for the festivity rain slowed their work. In spite of this, "juice" was available an hour or two before the start time.

Apparently there were many very long conductor spans as only every other pole had been set in place over the last few miles in order to meet the dancing deadline.

* * *

A CELEBRATION to recognize arrival of electricity in Wawota in the fall of 1955 was outstanding. I knew of no other anywhere in the province that compared. It was initiated by George Lincoln and his farm sign-up committee, the mayor and other town officials, appropriately called, "Lights up Wawota." Lincoln was the reeve of the rural municipality and a driving force in the community.

A few weeks after line construction had finished and the lines were energized, invitations went out to SPC personnel, Parcher, Crandell and I from the Yorkton Farm Branch office; Glen Gorham, the Yorkton commercial rep, who had handled the town's sign-up; Ted Rayner, the construction project foreman; Bill Gilbert, the Peebles DO; Gil Graves, the Grenfell district superintendent and his assistant, Gord Sim; and Roy Sarsfield, Charlie Smith and Jim Rutherford from Regina, with whom Lincoln had dealt before Garnet and I entered the picture. When the four of us arrived from Yorkton and checked in at the hotel, George welcomed us with a bottle of rye from his well-worn briefcase. Although he was a teetotaler he wanted to ensure our evening would be suitably toasted in our hotel rooms after the "dry" municipally sponsored festivities.

A civic dinner and reception for the invited guests, local dignitaries and the farm sign-up committee was followed by a program in the town hall packed to the rafters from both the townsfolk and the farm families. SPC staff sat in a reserved front row. Two huge colourful banners, made by the school children, hung across the back of the stage: "Lights Up Wawota" and "Thanks SPC." Lincoln, the homespun chairman, handled the program in his usual quiet, diplomatic, kind, yet forceful way. There were songs, skits, recitations and poems prepared by the children. A church choir sang, individual instrumentalists and singers performed, the operator of the town's little power plant was feted and presented with a plaque and a rocking chair in recognition of his many years of faithful community service, the farm sign-up committee was introduced, speeches came from the town's mayor, the school principal, the hospital board chairman and the Chamber of Commerce president, and finally, all the SPC people were introduced, explaining the role each had played to bring them electricity. Lincoln spoke nicely and at length about Parcher and I with whom he had the most contact. It was a warm, sincere, typical, unsophisticated, prairie social event, put on by folks wanting to celebrate this historic moment of their community's progress, just like they had done with the arrival of the steel, their first doctor, a paved main street, or a new school, grain elevator or hospital.

Charlie and Sarsfield, the most senior SPC management people there, were suitably impressed. Lincoln joined us for a short time afterwards where we toasted him. Then we partied.

This event had a big impact on me and reinforced that what I was engaged in was a worthwhile, in fact necessary, contribution to Saskatchewan society. I felt a great sense of pride being there that night.

* * *

THE FOLLOWING article, "Even Chickens Got Power," by Alvina Fellman of Creston, B.C., was published in the September 1999, issue of *The Saskatchewan Senior* tabloid publication. It tells another story of gratitude when electricity came to her part of rural Saskatchewan, a project organized by those of us working the Yorkton Area.

". . . We farmed in the Lake Edward district, east of Spalding. One memory stands out so much I will never forget it or stop appreciating it. Our district did not have electricity . . . and the running water we had was to run and carry it in and run to take it out. At last, enough people signed up and we were getting SaskPower. How excited we were. So the works got going. Power poles went up and everyone was wiring their houses. Some were wiring barns – even chicken houses!

It was December 1954, and we were told we would have power in time for our annual Christmas concert, which was a big event. Teachers and children worked hard. A nice big tree was put up and it had lights put on it in anticipation of the power coming.

On the day of the concert, the neighbours directly east of us got hooked up, but then it was quitting time and the men left. How I had the nerve – I will never know – but I phoned the Spalding Hotel and asked those poor men (who had worked all day in the cold) if they would come and hook the school up for the concert. I don't even recall what answer I got.

Our youngest son, who was two years old at the time, was so sick that my husband stayed home with him and our baby. The four older children and I trudged the very short distance to the school, feeling a bit down.

The school was full of parents and children . . . gas lamps doing their very best. Just before the concert was to begin, the whole school yard lit up! What cheers and clapping went on! There were the lights on the tree, what a beautiful sight. Those wonderful men from SaskPower had driven the six miles out, and using portable lights hooked up the power to the school. It was just awesome!

After a wonderful evening, the children and I started home – the children rushing to tell their dad the exciting news. Another surprise – our yard light flooded the whole place and every window in the house was lit up. What a thrill!

I don't know who the men were, but if by any chance they should read this and recognize this event, I would really like to hear from you. I would like to thank you again.

We, of course, have made many trips back to Saskatchewan and always go out to see the old farm. The school is long gone, just a sign that says Lake Edward School #2967. Nearby in the middle of the field stands the home we once owned. In my mind's eye, I see it as plain as 45 years ago . . . all lit up with the yard light showing us the way home. I did not see the granaries standing where the school was – I see the school with the lights shining like a beacon on a cold winter's night. I also see many hearts filled with gratitude.

I am thankful I can recall such precious memories. . . "

Other Electrifying Moments

THE FIRST FALL after the office decentralization to Yorkton, our program embraced a project a few miles west of the city: "Fonehill South." It included the tiny hamlet on the paved #10 Highway on its way to Willowbrook and Melville. Power line construction was to be completed in December. Organized the previous fall their wait had been unduly long. By mid-month bad weather made it doubtful that construction could be finished by Christmas. The committee chairman was the affable John Smerchinski, the proud owner of what the hand painted sign announced on his storefront and gas station, "Johns Genral Store." He phoned us often, came to our office many times, worried whether service would be available by Christmas day.

His concern, "My place is wired. I've bought a whole slew of Christmas presents for my wife and kids. All electric stuff – a train, a play oven and toys and a toaster and tea kettle and iron, electric range and a refrigerator. If we don't get the lights before Christmas my day will be ruined." Scotty Robb, the Yorkton DO, was well aware of John's predicament and said he would do what he could. As luck would have it Scotty and his assistant DO were able to get all the meters installed and service wires connected for all the farms and Fonehill customers on the 24th by mid-afternoon. When the lights came on at John's, Garnet, Scotty, his assistant and I were called to a Christmas eve celebration.

The look on his face as he toasted the four of us with a home-brew concoction of "chain-lightning" left no doubt as to his enormous pleasure. It was, in itself, a nice, warm Yuletide gift for us too. Thereafter, every time I stopped for gasoline at his pumps in Fonehill or met him in the city his smile still divulged his ongoing gratitude for our work.

* * *

AT EVERY organizing meeting I emphasized the benefits of electricity and the new way of life it would bring to my audience in both their farm yards and homes. Cattle could have a continuous supply of fresh water. Automatic feeders and lighting could make poultry more productive. Machine shops could be modernized, even with an arc welder. Homes could have running hot and cold water, refrigerators, freezers, electric ranges, automatic laundry equipment and the hand-cranked De Laval cream separators replaced with a motorized unit. Farm children would have better lighting with which to study. Wrestling matches could be watched on television, the same as what they saw in appliance store showrooms. Many of my listeners would stare quizzically at me, seriously suspecting the validity of my soothsaying. The look on their faces left little doubt of their disbelief of my predictions.

More than once when my wife and I were shopping in downtown Yorkton on a Saturday afternoon, a smiling farmer would approach, somehow expecting I would remember his face from the hundreds of others who sat and listened to me each year. A short conversation ensued and went something like, "Yur Anderson, right? Da elicrish man from da pawer carparation, ya?" I'd answer, "Yes." "Las yeer at da meetin in da skool, you vas right. Vee ar vatching da vrestlin on da telebision, jus lac you sad." It made me feel prophetic, that in just my lifetime my prognostications had come to pass. As we went on our way Jean would mutter to me, "What was that all about?" My reply: "Just another one of my happy customers."

* * *

IN MARCH, 1956, Charlie called out-of-the-blue and asked, "Dave, we're in a bit of a jam and I'm wondering if you could help with about a month's work near Maple Creek?" Doug Fee, the Swift Current farm field rep had been hospitalized with a back ailment. He handled the farm electrification work in this western part of the Regina Area. Although a long way from my Yorkton home, I wanted to help Charlie because of his decency towards me and my respect for him. I readily agreed.

There was urgency to complete this assignment. Power line construction work always started in the extreme south-west as frost came out of the ground earlier there than anywhere else in the province. The wood poles, conductor and line hardware had already been dropped off at several delivery points awaiting arrival of the construction crew. My task was to obtain from landowners free permission to cross their fields for two, new 25 kV power lines and offer electric service to those farmers whose buildings were within half a mile of the line's route, as we would connect them at the same time.

The line from Maple Creek to Leader, about 70 miles, would also serve the urban centres of Liebenthal, Mendham and Fox Valley. Leader had been supplied with electricity for many years by private interests, the system bought in 1929 by the Saskatchewan Power Commission. By 1956 that town's plant had been retired when it was served from a 25 kV line that followed #32 Highway through Cabri and Abbey from its source at Swift Current. The new line to Leader from Maple Creek would provide an alternate and more reliable supply to both communities and the surrounding areas. The other line of about 25 miles from Shaunavon to Eastend would serve the hamlet of Dollard. Both lines would end the life and long-time service of the small, relatively inefficient, diesel power plants in both Maple Creek and Eastend and, at the same time, make electricity available in future years to hundreds of farms and ranches on both sides of their respective routes.

There was still snow in the fields the first Monday morning I left Yorkton. I had to detour a mud slide over #10 Highway in the Qu'Appelle Valley, caused by spring runoff. By the time I stopped at Head Office for the project files and maps, then briefly visited Fee in the Swift Current hospital, continued on to Maple Creek and checked in at the town's historic hotel, it was too late for any work that day. I visited that evening with Mae and Chummy Everest, the Maple Creek District Electrical Superintendent, who was also responsible for running the power plant.

Next morning the RM secretary provided the names of landowners of each quarter section for the first few miles north to his municipality's border. As I worked my way further north similar calls were made to the RM offices in Fox Valley and Leader. The farmers I met were pleased to learn power would soon be available and offered little resistance to our plans, particularly those within the half mile of the line's route, most of whom paid the full AC construction charge to get service. I gave the names and addresses of absentee landowners to Head Office so explanatory letters of our intentions could be sent.

Word of my presence spread. As expected, pressure came from farmers whose buildings were just over the half mile limit. Often on return to my hotel in Maple Creek or Shaunavon, late and tired, two or three farmers were waiting to plead their case, "Couldn't you please serve me too?" "No, I'm afraid not this year," I had to say, but gave them petition forms and encouraged them to organize a group of their neighbours.

I worked from early morning to sunset for three and a half weeks to call on the more than 200 farmers to complete the easement clearances and sign-ups on both jobs. Poles were being planted by the construction crew north of Maple Creek before I had finished the Eastend job. After lunch each Friday I headed for home, reporting my progress to Head Office on the way through or on Monday morning as I headed west.

In June, 2002, I again drove both highways, the routes of those 25 kV lines, and reminisced about the focus of my task 46 years earlier. Both lines have been rebuilt, the 35 year lifespan of the original wood poles long since elapsed. The single-wire 13.8 kV farm tap-offs to each farm and ranch along their route have now had the poles removed and the conductor has been plowed underground. This innovative development started in the early 1980s and was, no doubt, appreciated by farmers, now relieved of the need to manoeuvre around poles and anchors with their now even bigger field machinery. Thousands of miles of single phase farm lines, about a third of the province, have been placed underground. Another advantage of this innovation is less exposure of the overhead lines to lightning and other natural or man-made causes of interruptions to a desired continuous supply of electrical service.

* * *

TED "BLONDIE" PHILIPCHUCK, the Balcarres DO was of Ukrainian descent and void of a blonde hair on his body. From 1950 to 1955 he did more to electrify individual farm customers than any other DO in our area. Throughout all the years of the farm electrification program it was our policy to connect any farm if it was within half a mile from a power line. We called these individual applications, "singles." There was only one proviso for their acceptance, the ability of the DO to either build the short power line himself or with the help of the farmer or by the three-man operations crew that came periodically to help each DO. Blondie consistently encouraged any and all farmers in his district who were within the half mile limit to apply for service. Applications from him always had the notation he could get it built. If a farmer applied directly to us, a call to Blondie was always in the affirmative. He operated his own self-help scheme for these single extensions, as in most cases the farmers helped with the grunt labour.

In one 1954 phone conversation with him came a request to see him the next time I was near Balcarres. When there he presented me with a completed farm application contract for a single, and a cheque for the full AC construction charge, both signed by Noel Pinay, a Plains Cree Indian. Pinay's farm buildings were just inside the boundary of the Peepeekissis Indian Reservation, north-east of Balcarres. Only a road allowance separated his yard from our power line. Blondie said,

"I can hang a transformer on an existing line pole and serve this farm with a new pole in the centre of his yard for the breaker and meter." I replied, "But we cannot accept a farm application contract without the registered owner's signature." In this case the land owner was the Government of Canada, Department of Indian Affairs. Pinay had signed the contract as the consumer, responsible for payment of the monthly energy account, but the line-space for the owners signature was conspicuously blank. Blondie was anxious to proceed, but I was uncertain.

Blondie suggested we meet Pinay, so we drove out. He seemed like a hard worker, had a decent set of buildings, good farm equipment and the land he cultivated, both on and off the reserve, appeared to have benefited from good field husbandry practices. Our line was close to the road allowance and I could easily have thrown a stone from the road and hit his barn. I was sympathetic and saw it as nothing more than running an extension cord from the edge of the reserve to his buildings. The lack of a landowner's signature seemed a pretty flimsy excuse to not serve him and his family.

We met Pinay, a nice guy, sensitive and polite with a strong attachment to the land, just like his ancestors. The three of us knew there wasn't a snowball's-chance-in-hell to get the white-faced Indian agent, let alone anyone else in that bureaucracy, to approve what seemed logical and reasonable. My thought was if he doesn't pay the monthly account we could salvage the above-ground material and only be out the labour costs, a small expense. SPC was taking no risk to serve this farmer and his family. As I would rather be right than consistent I disregarded our regulations and said, "Noel, sign here on the contract as the owner." And to Blondie, "We'll issue the work order for you to connect the first Indian farmer." Pinay smiled undemonstratively, shook my hand and quietly said, "Thanks, Dave." His demeanor plainly showed his gratitude.

I felt good about our serving SPC's first Indian farmer, a very tiny and insignificant step to compensate for the white man's oppression of these noble people. But I worried about the precedent and the trouble it might cause me if the Department of

Indian Affairs discovered what had been done and raised objections in Head Office. But I believed my defense would be a cake-walk. I checked periodically to ensure Pinay's monthly account was being paid. It always was. This matter never arose again.

Many boys of First Nations ancestry served in Canada's armed forces in the Second World War, 10 years earlier, some in the Navy, as did I, prepared, if called upon, to pay the supreme sacrifice for their country. But it would be six more years, in 1960, before Canada gave aboriginal people the right to vote in elections and two years after that, in 1962, before the fed's agreed Indian reserve residents could get electricity and begin to be treated the same as the white man, at least on this matter. Shameful government action, or inaction!

* * *

WHEN OUR ENGINEERS determined the routes of the rural lines for a project there were often tap-offs that followed what were called the blind-mile-line between those sections without a surveyed road allowance or a section's half-mile line to serve one or, in many cases, several other farmers who lived thataway. Where these mile or half-mile lines were well defined with fences our stakers accepted these divisions and placed the pole locations close to these fence lines. However, there were cases where no precise field demarcation existed because fences, if they had been there, had disappeared or were covered in drifted soil or the dividing line was obscured in heavy bush. In these instances the stakers had to measure from the iron survey stakes at the corners of the sections a mile or a half mile away. It was not unusual for them to uncover a field that had impinged several feet into the adjoining one. This discovery sometimes was the source of some disagreement between neighbours as one had, for years, cropped more land than to what he was entitled, to the extent that a rod, 16.5 feet, a half mile long amounts to an acre of farmland. I only heard of these cases well after the fact when talking to my committee men or a crew foreman. To my knowledge they were all settled amicably.

* * *

HARVEY MABSEN handled customer accounts in Head Office, and was responsible for the thousands of meter readings and the preparation of the bills and their collection each month. He was an old time Commission employee, went back to the days when customer power bills were prepared by hand. I first met Mabsen when I started with SPC in accounting in 1951 and when I was involved with both the Head Office social club and the Power Credit Union. He was one of those likable office characters, known within the company as "cut'em off Harvey." This moniker arose from the way he spat into his dictating machine sending memos to operations staff about customers with unpaid accounts. He would ask them to visit the delinquent, determine if there were legitimate grounds for the account to be unpaid, collect the money and, if he could not, to disconnect the service. As he concluded these emanations his emotions and body contortions entered into the act, climaxing the dictation in a loud irritated voice, "And if all else fails, cut'em off!" It was brash entertainment to witness this dramatic, yet unpleasant, aspect of his responsibilities.

Mabsen once called me about a farm account and said, "Dave, we have a farm customer near Sturgis that pays his bill promptly each month. But it's a mystery to us that for nearly two years, ever since the farmer got service, he has consumed exactly 54 kilowatt hours each month. Every self-reading meter card he sends us has precisely 54 more than the month before. It looks very suspicious. I don't want to bother the DO and wondered if you could diplomatically investigate sometime when you are up that way and let me know what you find out?"

The afternoon I drove to this farm the farmer was away, but his wife was home. She invited me into their home and I explained our curiosity about the consistency of their monthly electricity consumption. "Oh sure," she said. "My husband always watches the meter. He doesn't want to pay more than what you get for the $5 minimum. When it gets to 54 since the last time, the amount you get for $5, he goes to the pole in the yard and pulls the switch. We are in the dark until the next month starts when he turns the switch back on and then we have lights. Sometimes we are in the dark for nearly two weeks. Sometimes I am in the middle of washing or ironing the clothes and I have to finish by hand. The kids don't like it either when they don't have lights to study."

A bit astonished I explained the energy rate, a set of numbers I quoted from memory hundreds of times that are still in my head to this day.

Monthly rate for a 3 kVA transformer:
Service Charge - $1.00
first 45 kilowatt hours (kWhs) @ .08¢ each
next 45   "   "   "   @ .04¢   "
next 45   "   "   "   @ .03¢   "
all over 135   "   "   "   @ .02¢[3]   "
Minimum bill - $5

Then I carefully showed her how each kilowatt hour they used became cheaper as their consumption level rose. She was intrigued that for only $2 or $3 more than the $5 a month they were already paying that they could have lights every day for the entire month. I left her an energy rate card and wrote out calculations for several different kilowatt-hour consumption levels so she could explain them to her husband. She was delighted to have this information. I reported my findings to Mabsen and asked that he let me know what happened in subsequent months. He called sometime later, told me their monthly consumption had risen well above the minimum and the account continued to be paid promptly. She must have convinced him to never again touch the breaker-switch on their yard-pole.

How differently we each see the same things?

*   *   *

---

[3] In 1956, SPC's runoff electrical rate for farm and residential service was reduced to .01¢ per kWh.

MY WORST work faux pas which troubled, embarrassed and humbled me happened in 1953 in our Yorkton Farm Branch Office basement. I was meeting with the "Duff South" delegation who, on learning they would not be served until 1955, became upset and belligerent. I tried to explain that nearby groups had made application earlier, that farm densities, cattle population and land assessments of these other districts were higher and were important criteria for our project selection process. My explanations fell on deaf ears. In my frustration to justify our position and to convince them I displayed the maps on which this data was recorded. Erroneously I unfolded the map on which Cam Caswell, the Melville ag rep, had confidentially provided his opinions of the farming areas that were in his jurisdiction and they saw it before I could get it folded back up. Their district had been described with the word "POOR" emblazed on it.

Their anger peaked, demanding to know who had given them this label. Unable to cover my blunder I honestly admitted its source. Furious with me and with this information, they left in a huff. I called Charlie immediately. He admitted this would be an awkward situation, but understood how in the heat of discussion it had happened. He did not level any blame and told me to not worry. He would call Caswell. But before he was able to the group had driven the 30 miles to Melville and were in Caswell's office. He called me, angry and upset over my gaffe, leapt at me over the telephone like a cougar from the woods and took verbal bites from my backside. Then he called Charlie, who tried to assuage him. A few days later the Department of Agriculture officially objected to Head Office about my indiscretion. Charlie replied to their letter, apologized but supported me along this bumpy path. Little wonder I had such respect for Charlie Smith.

About 10 years later, when I was SPC's Public Relations Superintendent, after I had addressed a Melville Chamber of Commerce luncheon meeting, at which Cam Caswell attended, we spoke about that unfortunate incident. Apparently those farmers went out of their way to prove to him that they were not "poor." They became more active in 4H Club work, University of

Saskatchewan Extension activities and other programs which Caswell encouraged. While it took some years to live down his categorization of the district, the episode had been the source of other changes in this farming community. While I still felt awkward about the incident, I was glad there was an upside.

"Duff South" was electrified two years later, in 1955.

* * *

IN EARLY DECEMBER, 1956, Charlie seconded my help on two projects in northern Saskatchewan, near Flin Flon, Manitoba. The work was unrelated to my rural electrification work but are related here as they happened during my tenure in the farm job. I was pleased to be able to assist him and we caught a CNR train in Canora and travelled overnight via Hudson Bay Junction and The Pas, arriving in Flin Flon the next morning on a very cold day. Our purpose was to meet Hudson Bay Mining and Smelting Co., Limited (HBM&S) officials about their electric system supplying service to two nearby Saskatchewan communities. The first was the year-round occupied homes and the summer cottages at Denare Beach on Amisk Lake, south-west of Flin Flon. The second was at Sandy Bay, a native settlement that adjoined HBM&S's Island Falls Hydro Electric Power Station. This power plant, in Saskatchewan, north-west of Flin Flon on the Churchill River, produced electricity for the company mine and smelter, residents of the city of Flin Flon and the Town of Creighton, a Flin Flon satellite community, which was on the Saskatchewan side of the border.

HBM&S had, from the outset of building the dam and hydro station over 30 years earlier, denied service to anyone who lived outside the incorporated limits of Creighton and Flin Flon, alleging there was insufficient energy available to add more customers. This was ludicrous at Sandy Bay where the mostly native Indian people lived in the shadow of the power plant and were in the dark, even though many of the inhabitants worked at the station. Another obvious and outrageous betrayal of this argument was the hundreds of street lights that burned 24 hours a day in Flin Flon and Creighton. There was likely more electricity wasted in leaving those lights burning every day than would have been used by the citizens in the two communities with which we were concerned.

Charlie had researched the water licence granted to HBM&S by the Saskatchewan Department of Agriculture in the late 1920's, which had allowed the company to dam the Churchill River and build, own and operate the Island Falls station. One clause in the licence provided for one ninth of the energy produced to be available for the benefit of Saskatchewan people.

When we met their vice president of operations Charlie told him of the demands for electric service we were receiving from Saskatchewan people living in Sandy Bay and at Denare Beach. The company official immediately reiterated the standard argument HBM&S had used for years that there was no surplus energy to serve more customers.

Charlie pulled a copy of the 35 year-old water licence from his worn briefcase, leafed it open to the page containing the clause in question, read it, then quietly said, "We will be exercising those rights for the people living in Saskatchewan at Sandy Bay and Denare Beach which, when served, plus the town of Creighton, will still be a long way from one ninth of the plant's output." The company official stared at us in disbelief, then Charlie continued, "Over the next day or two we will take the initial steps to satisfy those needs. SPC will own, operate and maintain the two extensions and will purchase the energy in bulk from HBM&S at a price we will negotiate with you later." The company VP was annoyed their indecent and illegal act had imploded, that of denying Saskatchewan residents a service to which they were entitled. I could not believe this international, credible, industrial conglomerate had, without any virtue of honesty, intentionally perjured itself for so many years with local people, many of whom were their own employees, who were in need of their services which they were legally bound to provide. The integrity of big business?

The next day we rented a car and I took Charlie to the airport where he took a chartered flight to Sandy Bay to meet the

Indian reserve people and assess the line layouts needed to serve them. I went to Denare Beach and met the Local Improvement District staff who administered the resort. I got a list of all property owners and spent the day field-checking the community with a map, identifying the size and location of the homes, cottages, store and other commercial customers which included the DNR fish plant. The latter had its own generation to process, freeze and store locally caught fish, which included large quantities of sturgeon, caught at the Sturgeon Weir a few miles south. These huge frozen and glazed fish, two to each wooden crate, were exported by the airplane load to Chicago's kosher market.

At a public meeting the next evening, advertised by word of mouth amongst the cottage owners and permanent residents, we explained our plans for Denare Beach. The people were elated. When back in Yorkton I had cost estimates prepared for the line from Creighton, plus the distribution and street light system for the resort. Construction charges were quoted to the customers by mail and the sign-up was handled by the municipal staff.

In the summer of 1957 Denare Beach was served. After construction was finished a local reception and dinner was held to which Cass-Beggs, Charlie and I were invited. We attended, travelling by charter aircraft from Saskatoon. Delays attributable to bureaucratic red-tape in the Federal Department of Indian Affairs meant service to Sandy Bay did not happen until 1958.

* * *

JUST BEFORE Christmas of 1956 an RCMP investigation uncovered an illicit home-brew operation in a vacant farm house a few miles north-west of Yorkton. The unlawful still was in the earthen dugout basement where the rascals had illegally connected the home's electrical service to the unused, but still connected transformer on the pole in the farmyard, a dangerous and risky venture. As there was no meter on the pole to record consumption they were stealing the juice for brewing and distilling their hooch.

A substantial quantity of the finished product, aging in barrels, was awaiting bottling, distribution and sale. Under the law it had to be destroyed so the uniformed officers carried the clear liquid, by the pailful, from the basement and dumped it outside in the yard. The brew found a fissure in the frozen earth and drained back into the dugout basement making the dirt floor a quagmire. A neighbour, Ewen Gray, an RCMP sergeant, was part of the investigative force and got his shiny brown boots and feet soaked in the concoction in the freezing temperatures. He later came down with a severe cold. His unsympathetic neighbours teased him about the event and complained he hadn't even brought a pailful home to treat his affliction or, more importantly, us.

Out of curiosity I later went to the farm. It looked doubtful the house would ever be occupied so I had the operations staff remove the transformer from the yard pole to eliminate any temptation for reoccurrence of the felony. While the RCMP charged and convicted those responsible SPC made no attempt to prosecute or to recover the value of the stolen energy with the "jimmied" wiring as it would be impossible to identify the culprit electrician from the rest of the gang.

Theft of electricity by dishonest customers, including farmers, is an ongoing, serious problem for power utilities.

* * *

MY DEEP respect for the farming community, with whom I had worked so closely was seriously bruised in 1969. I was, at the time, SPC's Electric System Business Division Manager, several rungs up my personal ladder from my farm field rep days. One of my responsibilities was the administration of electric rates of energy applied to all SPC customers to ensure the design of the rates produced adequate revenue to offset the costs of serving those on that particular tariff. My staff doing this work was headed by Jim Rutherford, the former Regina Area farm field rep, who conducted the organizing meeting at Lajord in 1952, early in my farm electrification work.

A serious, ongoing, nagging problem for utilities, both electric and natural gas, is "energy diversion," the innocuous name it is called within industry ranks. Some of it happened because of internal management errors and oversights, but a lot was what I preferred to call, just plain, old, ordinary "theft." During my tenure as business manager my staff's initiatives recovered millions of dollars, revenue that was rightfully the Corporation's. This function was an internal auditing, behind-the-scenes, almost clandestine operation, one that received no publicity within or outside the company. It was one of those plodding day-by-day sort of business functions, examining suspicious cases, one after another.

Lost revenues were the result of accounting system errors which under-billed customers the amounts owing SPC. There was careless meter readings by inexperienced and inept staff who did not receive adequate training to alert them to meters that were inappropriate for the load being served or were functioning improperly. DOs and other field operating staff often did not receive the necessary training to give them the expertise to properly assign the correct rate to customers or to administer the rates already applied to existing consumers. And then there were the dishonest customers, the do-it-yourselfers and electrical trades people, knowledgeable about wiring and how to covertly divert energy. Finally, there were those customers, a level of humankind that exists in every segment of society, from the petty pilferers in shops and stores to the more sophisticated break and entry artists and on up to the embezzlers in the highest echelons of business and industry, so why would the farming community be an exception and not have its share of these nefarious people?

SPC's 1969 annual report, one of the most important of the Corporation's official documents, submitted to the governing Liberal government of the day and the citizens of the province, stated, "The average annual consumption per farm customer increased sharply over 1968 to 8,250 kWh per farm. This increase of almost 15% appears to be the result of increasing diversity and mechanization of Saskatchewan farms." This statement was, for the most part, hogwash! The real reason could not be stated, which adds credibility to the expression, "You cannot always believe what you read," even in official company publications. At the time the report was being prepared, in January of 1970, SPC's senior management were yet to be alerted to what follows and were unaware of the truth on this matter so could not, in any way, be accused of intentionally misleading their legislative bosses. While average farm usage did grow in 1969 from the previous year, perhaps six or maybe even seven per cent, the larger part of the increase was our recovery of revenue from farmers who deliberately stole electricity and avoided paying for it.

At the start of the rural electrification program SPC asked farmers to read their electric meters monthly on a company supplied postage-paid, mail-in card. On arrival in head office they were processed, a bill was prepared and mailed.. Sometime later this was changed to a three month reading schedule. Farm meter dials displayed four numbers up to a maximum of 9,999 kWhs when the dials automatically reset themselves to zero, then continued to register the energy being used. These four dial meters were adequate when electrical service was first provided and consumption levels were relatively low. However, with the increased popularity of electric space heating in farm homes, the availability of a growing number of household appliances and equipment plus the more intense agricultural operations in large barns with enclosed environments for livestock and poultry, farm electrical usage rose significantly.

As the average annual farm consumption of electricity in 1969 was over 8,000 kWhs per customer, approaching the number before the meter recycled, it was blatantly obvious in hindsight that thousands of farm meters recorded yearly levels of consumption that were in excess of the meter's capability before it reset at zero. This was an open invitation to those less exemplary than the vast majority. SPC's field operations people, DOs, district superintendents, area managers and, at the very least, the Head Office metering engineering staff should have been on to this problem long before with revised metering standards that would have prevented what will be described here.

Each fall it was standard practice for DOs to inspect every rural electrical installation within their jurisdiction, to verify records of the serial numbers of transformers, pole mounted circuit breakers and meters, to check the power line and hardware on the transformer pole in the farmyard to ensure all equipment, ours and the farmers, was in a safe and reliable condition and, where deficiencies were discovered, to make the needed corrections.

The Yorkton DO, in the process of making inspections in the fall of 1968 to his farm customers, became suspicious over the consumption levels of four farms in his district, near Rhein. They all were served with large transformers but the amounts of electricity they used seemed abnormally low considering the size of our electrical installation on each yard pole. He uncovered a devious scheme when his curiosity prompted his investigation. He discovered the meter reading numbers sent by the farmers on their self-reading cards every three months were deliberately less than the actual readings displayed on the meters. He verified his suspicions with monthly trips to each farm and before Christmas, when he was certain of his facts he and his boss, Eddy Edison, the Yorkton district superintendent, called me, explained what they had found and asked, "What action should we take?" I asked for a written report and told them as soon as it was received I would call them. I then consulted with my counterpart Don J. Anderson, Electrical Distribution Manager.

Their report confirmed premeditated theft of electricity. Company lawyers told Don and I we should take our lumps and forget about any retribution. To take legal action and press charges against the four farmers would be difficult to prove in a court of law as there was no hard evidence, only circumstantial, and each case would have to go through this convoluted process on its own, likely with questionable results in our favour and legal costs of more than we would ever retrieve. Don and I considered the issue for a day or two, then disregarding our solicitors advice, decided to take things into our own hands, feeling the perpetrators should not get off scot-free at the expense of honest customers. Besides, we well knew there would never be the political will by senior management, and certainly not by our political masters, to legally confront these farmers and, if we pursued the matter legally SPC would have to admit carelessness, that we had improperly minded our store and left the cash register unattended. So on our own we collaborated on another plan of action, one that we kept to ourselves and did not report immediately to our superiors.

Rather than determine how long each culprit had taken advantage of us we asked Edison and his DO to change the meters on those farms to five dial units, then we arbitrarily sent a bill to each of the four for 10,000 kWhs, one full cycle of the meter, at the then farm rate runoff price of 0.02.2¢ per kWh, which amounted to $220. All were promptly paid which, in itself, was an admission of their guilt.

Concerned the thievery might be more widespread than the initial discovery and, knowing that popular wisdom has it that where there is smoke there is fire, my staff explored further. From the computerized billing records the consumption level of every farm in the province was compared with the size of the transformer with which it was served. When anomalies arose from normal consumption levels that could be expected from the transformer size that served the farm, the computer spewed out lists of suspicious accounts. I did not want to believe the problem was as significant as that list seemed to indicate. But I was about to get my eyes opened.

Don Anderson then instructed all his field staff to replace the old four dial meters with new five dial units on those large transformer installations and send us the readings from the new meters every month for the next three or four months. The frequency of Don's staff in the farm yards of the suspects after the new meters were in place undoubtedly alerted them that the jig was up. It was readily obvious if theft had occurred when these new readings were compared to the earlier consumption levels sent to us by farmers on their self-reading meter cards. As 1969 unfolded scores were sent bills for $220. Every single one was quietly paid without a comment or complaint. By year-end thousands of four dial meters were changed on all farms with large electrical consumption which put an end to the shenanigans.

These invoices for 10,000 kWhs plus the correct and higher quarterly consumption recorded on the new five dial meters, which produced at least another 10,000 kWhs or more that year from the affected customers, was the truth behind the abnormal increase in average electricity use per farm in 1969 over 1968 as reported in that year's annual report. A quick glance at a graph of the average annual electrical consumption levels of farms over several years clearly showed the anomaly of the 1969 statistic, unexplainable except for theft.

The thefts were manipulated this way. Three months after the DOs routine check of the farm installation in the fall the farmer mailed the meter card with a number that was 2,500 kWhs less than the actual consumption recorded by the meter. Six months later, the second reading sent in was 5,000 kWhs less. Nine months later the third meter reading showed an accumulated consumption that was 7,500 kWhs less than the actual reading. Then the next fall when the DO made his usual inspection the four numbers displayed on the meter looked about right but, in fact, were at least 10,000 kWhs lower than had actually been used in the previous twelve months. This shortfall on some high consumption farms could well have been multiples of this number.

There were two noteworthy cases. One, a sitting MLA of the NDP government, a farmer located in the Saskatoon area of the Distribution Division. Bob Tessier, the Saskatoon area manager alerted me, "How should we handle this one?" My response: "The same as all the rest." The other was a large grain farmer south of Craik, who happened to be the son of a friend of my mother-in-law. A 25 kV transmission line passed close by his farm site. We had earlier given considerable advice on a three phase electrical service to supply his large workshop and farm motor loads as well as technical advice to efficiently move stored grain. It was a bit ironic that our tours of farmers and other groups to the Outlook irrigation project always stopped to show off this mechanized farmyard and electrically heated home. We always had given him and his wife a gift for their trouble to accommodate us.

At the outset of our discovery of these felonious practices I was worried they might be of epidemic proportions. However, we came to believe the scheme was spread by beer parlour talk or from conversations around the elevator office stove or at livestock or other agricultural meetings. When one farmer got away with it, they likely could not resist boasting to a relative, friend or neighbour. We surmised the temptation was simply too great to beat big business and SPC was certainly one of the province's biggest. There were enough corrupt politicians around at every level, so the thoughts or tendencies of many people existed: "Let's beat them." So why not thievery against an organization politicians controlled? I was concerned there might have been a concerted effort by some individual or organization to market the scheme, but evidence of this never surfaced.

Obviously, by a long shot, the vast majority of farmers were not involved, but those that cheated distressed me and brought out a level of anger I made no bones about. Much like rust infecting a field of wheat this felonious scheme spread amongst farmers, that group of society whom I naively believed had such a high level of integrity, a luxury the thieves could obviously not afford. They had betrayed themselves and took advantage of SPC's sloppy business practices. Their profound gratitude when we worked so hard to bring them service initially had long since been forgotten, sad commentary on the value changes of a more greedy, self-centered, materialistic minded society.

Had this matter become public knowledge I can barely imagine the extent of the furor that would have erupted in that year's review of the Corporation's annual report in the Crown Corporation Committee meetings at the provincial Legislature. It would have been pretty juicy stuff any Opposition member would have gladly given his eye teeth to expose against first, the dishonest politician likely resulting in his resignation and, second, the Corporation's board chairman, senior management and Board of Directors. Instead there was glib acceptance of what was stated in the 1969 annual report.

Dave Furlong, SPC's President, appointed by Premier W. Ross Thatcher in 1964, was both disturbed and delighted with our report when he learned the details of this devious practice and our strategies for its resolution. The financial gains for the Corporation were all profit, every cent of it.

* * *

THE MARKETING of electricity, aimed at increasing its use by customers, was an important part of the sales division's mandate, to maximize the Corporation's return on its investments through greater use of the installed assets. Roy Sarsfield, the division's manager, was a resourceful marketer, never short of promotional ideas and schemes.

Two major campaign slogans that extensively promoted greater use of energy in advertising at that time were, "Live Better Electrically" and "Farm Better Electrically."

To spearhead his marketing plans in 1953 Sarsfield created a Load Development Department. George Busse, a University of Saskatchewan graduate Agrologist became its manager. He had formerly worked for the provincial Department of Agriculture.

An adequate wiring supervisor, Rafe Walker, was already on staff. He worked with farmers from the earliest days of the farm electrification program on self-help electrical wiring education when farmers displayed a strong interest to do this work themselves in their buildings. In 1955 Rafe retired and was replaced by Al Fraser, a qualified electrical contractor. Also hired about the same time was Doug Wilde, an agricultural engineer, whose task was to provide farmers with technical information about their use of energy in the farmyard. Lillian Vigrass, a home economist, who had been raised in Pathlow, near Melfort, was employed in 1956, timing that coincided with a reduction in electricity rates which gave a run-off or end-rate of one cent per kWh for all residential and farm customers. This prompted

Sarsfield and Busse to coin the trade-name, "Penny Powers," which all of the many SPC home economists were called thereafter. This moniker became a household name with women, and a significant part of the male population across the province, even when the runoff rate later increased well above the penny mark. In 2001, Lillian was inducted in Saskatchewan's Agricultural Hall Of Fame in recognition for her educational work to the farming community during the rural electrification program.

* * *

ONE OF THE EARLY promotional efforts with farmers as the target audience was distribution of the monthly magazine, *Electricity on the Farm*, sent for free to every electrified farm on SPC's system. Farm customers in Alberta and Manitoba also received the publication, provided by their power companies. The magazine featured articles about the farming community, trends in agriculture, how-to stories and photographs about resourceful and novel uses of electricity in their homes and farm yards. As the magazine was produced in the United States many of its articles were not applicable to Canadian prairie agriculture. In about 1954, Sarsfield felt a locally produced publication would be more suitable and, as well, could probably be done cheaper than purchasing the American publication. He garnered support for his idea from his counterparts at Manitoba Hydro, Alberta Power and Canadian Utilities. The combined number of electrified farm customers served by the four utilities in the three prairie provinces had then grown to about 100,000, sufficient for a viable, made-in-Canada periodical that would have better customer acceptance. Sarsfield's astuteness led to the publication being produced in Saskatchewan.

Bill Bradley, of Regina's Bradley Publications was engaged to produce a tabloid sized newsprint paper, 10 issues a year, called *Farm Light And Power* (*FL&P*.) With the support and cooperation of the participating utilities the publication was sent to Canadian prairie farm homes for over 40 years until the mid-90s, each issue filled with articles about the wise and efficient use of electricity along with ads from suppliers of electrical and farm equipment, and other agribusinesses.

Each of the four utilities had exclusive use of *FL&P*'s two centre-fold pages for ads, articles and news items which, when distributed in their respective service areas, gave the paper a provincial flavour. In exchange for this exclusive editorial and advertising space each utility made available to Bradley the mailing lists of all farm customers on their respective billing systems. The utilities paid for a share of the mailing costs, an item that came under much discussion as postage rates increased over time. Bradley sold other advertising and retained this revenue to cover the paper's production and printing cost, which included his staff, paid farm journalists for freelance articles and for other writers and columnists.

Initially much of the Corporation's input of printed material and information for the paper's centre spread originated the Public Relations Department. However as more and more farm customers were added to the system and as more load development specialists with energy use expertise came on staff, the preparation of more pertinent articles and data came from these people. The University of Saskatchewan's College of Agriculture and its Extension Department were regular contributors of technical information on farm management practices, news about animal and field husbandry issues and reports of current agricultural research work. Messages about electrical safety, both low and high voltage, were regular features. Bradley's familiarity with the farming community and the expertise offered by his staff and other journalists enabled *FL&P* to be considered by farm readers as a credible voice on the province's agricultural scene.

FL&P's editorial content was controlled by a board of governors including Bradley and one from each of the four participating utilities. Roy Sarsfield represented SPC from the paper's inception until 1967, then for the next 10 years I assumed the role coincidental with my appointment to replace Sarsfield. This was several rungs up my success ladder from my original junior accounting clerk position. During my 10 year stint on *FL&P*'s board of governors I was their chairman for two, two-year terms.

It was the paper's editorial policy from the outset to prohibit any political or propane gas publicity and advertising on any of its pages. The ban on propane was because it was a competing energy source with electricity for water and space heating purposes on farms. Political ads were restricted to keep the paper apolitical. Maintenance of this latter policy was not easy at times. When Jack Messer was Minister of Agriculture in Premier Blakeney's cabinet and, as well, Chairman of SPC's Board of Directors in the 1970s, he well understood the reputation and acceptance of *FL&P* in Saskatchewan's farming community. When I was board chairman, there were three awkward confrontations with Messer to defend the paper's policy after Bradley had declined preelection ads from the New Democratic Party, the renamed successor of the CCF, and the province's Department of Agriculture. I won all three bouts, but came out of them a bit bashed, bruised and beat-up.

When the rates for electricity started to increase and energy costs became a concern for farmers SPC withdrew from the blatant promotion of greater use of electricity to encourage customers to more prudently consume energy. So the paper became less of a promotional vehicle to one that encouraged energy conservation, its wise and more efficient use. In 1975, during one of my stints as *FL&P*'s Board Chairman, a prolonged postal strike placed the paper in financial jeopardy, but we limped through this crisis, as well as a number of postal rate increases that also threatened its viability.

* * *

AL FRASER, SPC's Adequate Wiring Supervisor, assisted customers with wiring problems, including meeting with farm groups about to receive electrical service for the first time and who intended to wire their own buildings. SPC's 1954 annual report records there were 54 such meetings held that year. These were the same farm groups the farm field reps had organized a few months earlier.

Usually held in the same locale as the earlier organizing meeting, he often combined several projects when they were in close proximity. At these sessions the electrical wiring code standards were explained: how to obtain an owner's wiring permit, the importance of electrical safety and the promotion of more adequate wiring in excess of the minimum code requirements where this would be of greater convenience or benefit.

SPC purchased from the Electrical Inspection Branch of the province's Department of Labour, a supply of two books: *The Canadian Standards Association – Electrical Code* and *Wiring Simplified.* Fraser resold them at our cost at these sessions, hundreds of copies annually. With regularity the Department of Labour's budget for the books was exhausted well before the government's March 31 fiscal year-end when they were unable to supply our needs if we had not ensured a supply beforehand. This was bewildering. Since they sold them to us at a profit, it seemed logical to sell more books to make more profit. However, government bureaucrats don't think this way. Book costs were a departmental expense. Book sales revenue went to general revenue, not as an offset of their cost. Absurd and dumb, then and now.

In 1962 when I was SPC's Public Relation's Manager, one of my staff writers was a chap by the name of Cy Knight. An assignment given him was a photo-story about Al Fraser's work for use in SPC publicity and also in *FL&P*. Knight accompanied Fraser on an overnight jaunt to an evening wiring meeting with farmers near Melfort. Knight took photographs and made notes about the meeting. Fraser was meticulous about the care of his company assigned car and to his credit treated it like his own, always keeping it clean and in good repair.

Returning to Regina the next day Knight asked if he could drive. Fraser agreed, unaware Knight did not have a driver's licence and in fact, had repeatedly failed driving tests. All went well on this straight stretch of highway with Knight at the wheel until they approached a maintenance crew's barricade blocking their side of the road to await oncoming traffic. Knight slowed, then his imperfect inexperience took over. The car suddenly accelerated, collided and glanced off the barricade, veered into and through the ditch, then got tangled-up with a barbwire fence. The runaway dragster lugged its menagerie of wire and posts through a farmer's crop until it turned and attacked the fence yet again as it headed back towards the ditch and the highway. At the ditch's edge the car mounted a rock pile, stalled, hung-up on the rocks and became immovable. As well as considerable vehicle damage, the farmer had to be paid for the crop loss and fence mutilation. The car was towed to a garage and patched-up, sufficient to get them home before the day was out.

Fraser was more than a little indignant over the accident, and rightly so. Company procedures required a driver's report of the incident, so Knight obliged, inimitably and creatively in defense of his ineptitude. Fraser, as the vehicle assignee and witness to the affair, refused to co-sign the epistle when Knight delivered it to him. Fraser angrily brought it to me. Knight's description of the escapade fantasized about some mystical force that caused the car's acceleration when danger raced towards them on the highway. This sudden siege contributed to the barricade collision, then the fence entrapped them as the driver tried to escape through the field and crop, but a rock pile intervened where his dispatch professed, "and then the driver lost control."

I signed the accident report. Fraser was miffed over the whole deal. Knight resigned from SPC soon after, but the report of his classic caper was the source of much jocularity long after his departure.

* * *

HIGH VOLTAGE electrical safety was mentioned, but only briefly, at each of my organizing meetings. I dealt more with the transformer pole's location in the farmyard and its proximity to buildings and water wells, than about the height of farm equipment, to not raise aluminum irrigation pipes or locate granaries, hay stacks and bale piles under power lines. This topic came under more focused discussion at Al Fraser's wiring meetings after the sign-ups were finished and before the lines were built.

With the profusion of thousands of miles of new rural power lines throughout the countrysides each year public safety was a subject of serious concern to SPC. There were many tragic high-voltage public accidents, several were fatal, arising from activities around the farm yard's transformer pole and in fields where high farm equipment and other activities, like the placement of hay stacks took place. Farmers, family members and their hired hands were critically burned or disfigured or electrocuted when they contacted live overhead wires when hay was inadvertently stacked too high under a line, or when aluminum irrigation pipes were upraised to rid them of rodents and debris, or when cultivators and grain augers were in their travel modes and raised too high to safely pass underneath.

We had one case of a farmer painting his yard pole, and setting the paint can on top of the transformer. As he reached around it he inadvertently touched the live line and, tragically, was instantly electrocuted and fell to the ground.

Mishaps involved children too, when they climbed into fenced community substations to retrieve a wayward ball or flew kites in fields which became entangled in overhead lines then used long sticks or pipes to dislodge them. Building movers recklessly perched on the roofs of structures they were hauling and dangerously hand lifted energized rural lines as they passed under. Well diggers, overhead crane and high boom operators repeatedly had miserable mishaps and were perhaps the most negligent group with which the Corporation had to deal.

To help educate Saskatchewan citizens to be alert to the dangers associated with high voltage facilities SPC developed a public safety campaign around the theme, "Look Up And Live." A sharply focused media campaign promoted this slogan and related messages with articles and paid media advertisements that appeared repeatedly in all provincial weekly and daily newspapers, spots aired on every radio station and one minute commercials shown on each television station. These messages were highlighted in *FL&P*. Their frequency was intensified during seeding, haying and harvest, times when farm field activities and the potential for accidents was the greatest.

Public demonstrations of high-voltage electrical safety using table-top scaled down models were presented by SPC's safety officers to children in schools, to farmers at farm field days and to the general public at exhibitions like Regina's Agribition and at rural fairs.

In spite of all our efforts they seemed hopelessly inadequate when mishaps occurred.

* * *

IN ABOUT 1954 George Busse launched a series of Farm Electrification Educational Field Days that were held in the summer months between seeding and harvest. A semi-trailer van was packed with electrically driven farm equipment and appliances. With the help of the ag reps and local agricultural committees, electrified farms that were located on good municipal roads or provincial highways were selected for the event. *FL&P* featured their locations and other details of all the field days and, as well, local weekly newspapers and radio farm programs advertised each one when they were in their coverage areas.

Busse and Doug Wilde led these field days, supported by the local ag rep and the two farm field reps in whose area the event was staged. After 1956 Penny Powers was part of the field day agenda to disseminate information targeted to the farm women. They were often assisted by summer-employed undergraduate home economists, or on occasion, a home economist from the U of S Extension Department.

On the eve of each event the local DO connected the tractor-trailer's wiring to the selected farm's yard pole. Next morning we unloaded the van, a network of extension cords fanned out to tables set up around the farmyard and in outbuildings to display feed grinders, arc welders, grain augers, various kinds of electric motors, water pressure systems, power saws, drill presses, grindstones, hand drills, sanders and a portable light plant, all with associated literature. When it was unloaded the front of the trailer became a demonstration kitchen with built-in cupboards, a refrigerator, electric range, washing machine, clothes dryer and a deep-freeze. When the crowds arrived after lunch the men were directed one way, the ladies another. The farmers moved from display to display for talks, to watch demonstrations, to use and inspect the tools and gear and to have their questions answered. Educational and promotional films were shown in a darkened part of the barn. Electrical safety was always stressed.

The empty trailer, converted into a mini theatre, had portable benches which faced the model kitchen where Penny Powers gave demonstrations to the ladies about the selection, care and use of household electrical equipment and appliances. If large numbers of women attended they were split into two groups and the other taken into the farm home for a similar program.

Sometimes the local agricultural committee arranged a guest speaker from the University of Saskatchewan's College of Agriculture to talk on some aspect of field or animal husbandry or farm management or machinery developments. These speakers, usually well known and respected people in the farming community, helped attract large audiences.

At day's end the gear was repacked, loaded in the trailer and readied to move to the next venue. These were long days of hard work without the aid of a hydraulic-lift-loader and, as much of the equipment was weighty, our backs became the "beasts of burden." We taunted Busse, "If it was heavy, we had it." Often after we packed up, the farmer whose yard we used had a barbecue or potluck dinner sponsored by the agricultural committee, or the group would host us at a nearby restaurant before we moved on.

Early one evening in 1956 after a field day north of Wadena, we headed for the next venue near Melfort. When our convoy of three cars and tractor-trailer arrived at the Barrier River, north of Archerwill, Busse stopped the entourage and announced, "I think we should take a break and have a fresh fish feed." Good idea! He led us a few miles east of the highway to a bridge over the river from which he had fished previously. While Busse cast his line, we relaxed with a beer from the van's kitchen display refrigerator. Soon several pickerel fillets appeared, the portable light plant was started, the electric fry pan was plugged in and with butter from the fridge and seasonings from the tractor trailer's cupboards Penny Powers prepared a memorable repast.

After we returned to the highway about dusk and headed north to Tisdale my station wagon's transmission conked-out. I could only move slowly in first gear. The van and one car with the girls went ahead, the other car shepherded Jessop and I to a late arrival in Tisdale, where I left the car, doors unlocked and keys in the ignition, at the Ford garage with a note and a Field Order requisition under the windshield wiper for the mechanics to repair it the next day.

At a nearby cafe we ordered dinner, preceded by an appetizer of many "prairie oysters." The Chinese owner was a bit distressed as he watched us devour what he thought were unreasonable quantities of soda crackers buttered around the edges to hold some tomato catchup, topped with a dollop of HP sauce. While rather sad imitations of the real thing they were better than nothing. We paid him extra for the hors d'oeuvres. It was well after midnight before we continued west, arrived at Melfort and checked in for the night at the Ozark Hotel.

* * *

IN THE EARLY DAYS of the province's electrification power failures were commonplace and yet folks managed to persevere through these inconvenient times with little difficulty. These outages were caused by faulty or failed equipment in substations, human error, overloaded lines or power stations, storm damaged lines from lightning or fallen objects. However, as the high-voltage transmission grid was expanded it connected all power stations as if their generators were connected to one common shaft. These improvements and the interconnected lines then enabled large areas to be served from two or more sources when one was in trouble. So outages became less frequent and when they did occur were usually of shorter duration. Better communication systems for SPC's operations staff helped immeasurably. High voltage equipment improvements also contributed, like fusing devices on the single-wire 13.8 kV farm lines which allowed direct lightning strikes on poles and conductors to be only a blink in the lights as opposed to total darkness, perhaps for hours, before their invention.

As the reliability of SPC's electrical service increased the ability of customers to cope when it failed correspondingly decreased. Coal and wood ranges vanished in favour of electric ranges, thermostatically controlled heating systems replaced manually fed furnaces. Hand drawn wells succumbed to automatic water systems. These life style changes concerned us, particularly in the rural areas. Some bitter lessons were learned by those without some source of emergency standby preparedness in wintertime, like a stand-alone electric generator, a basement wood stove, a fireplace or other emergency provisions.

When I was Business Manager for SPC in the late 1960s there was a prolonged wintertime outage, more than a day, in the province's south-west. Many communities and hundreds of farm customers were without electrical service while our operations staff waited for the blizzard to blow itself out. When the storm abated the DOs laboured long hours to repair damaged equipment, downed lines and replace blown fuses on farm lines and in town substations. As one of them drove by a farm home that had been affected, the occupants of which he was acquainted, he noticed no sign of human activity around the house.

His knock at the door went unanswered. Inside he found the family in the basement huddled in blankets around a kerosene camp stove in a semi-conscious state but luckily all were alive. Their new house was electrically heated, tightly built and well insulated without a chimney. With no easy access for fresh air to enter the home the stove had used-up the available oxygen and had produced carbon monoxide, an odourless, toxic gas, which caused the emergency. Help was summoned, all survived.

This was a rude wake-up call for the Corporation. There was need to publicize emergency preparedness by customers when outages occurred, a quietly growing need requiring education and publicity, which was done repeatedly in *FL&P* and at presentations to farm groups as the opportunities arose.

* * *

SPC STARTED to promote the installation of farm water and sewage systems in 1958, spearheaded by Doug Wilde and assisted by Gord Seel, Doug Fee and other farm field reps. With adequate supplies of potable water on farms their electrical consumption would rise substantially, as power would run water pumps on pressure systems and heat water for both domestic use and in livestock waterers. As well, a water supply in the home meant modern kitchens, laundry rooms and bathrooms. To demonstrate these systems several farms around the the province were selected much like what had been done for the farm field days. With each family's concurrence a full water and sewage system was installed with the farmer paying the material costs and SPC covering the labour expenses. When the installation was completed, except for back-filling the water and sewer line trenches, a field day was widely promoted with advertising on radio farm broadcasts, in local weekly newspapers and in *FL&P*.

One such demonstration was on Teofil Ikert's farm, south of Wapella. Ikert had been a committee chairman of mine in 1957 when "Wapella South" was electrified. His water well was upgraded, six-foot deep trenches were dug and plastic pipe was laid to the house and barns, a sewage disposal septic field was built, a pressure system was installed, water heaters were connected to supply the kitchen and a new bathroom. The field day attracted hundreds on that cold November day. But the weather did not diminish the enthusiasm and joviality that night after the crowds left and the traditional fun ceremony of burning the outhouse down was celebrated with dinner prepared by a grateful family. Invited to attend the Ikert event, I was not then involved with farm electrification and this particular promotion. I mention it here as an indication of the ingenuity of SPC's sales division staff to assist the farming community to an easier way of life while at the same time increasing the consumption of electricity.

The provincial government decided this successful program should be coordinated with their parallel activity of water systems for villages and hamlets. So, with misgiving and some ill-will, SPC backed away from this particular farm promotion after it had been developed into a smooth and workable format.

* * *

IN 1959, Roy Sarsfield moved me to Regina, to work with him on special assignments, as assistant to the sales manager. As my family and I prepared to leave Yorkton I had a not so unusual Saturday afternoon backyard visit from Ewen Gray, my RCMP neighbour, except this time he looked uncomfortable and sheepish, strangely apologized for the visit. His superiors had insisted he see me as they knew I'd been exposed many times to illegal alcoholic spirits, maybe even knew of places of its manufacture. They wanted to curb this popular pastime and Gray asked me, "All you need to do is make a list of names, addresses and locations and give it to me. The force will protect you as the source of the information." Their assumption that I had some knowledge of the matter was dead-on, but I told Gray, "You're right, Ewen. I've had drinks of home-brew from many of my farm sign-up committee members and have been offered this brand of refreshment many times in the homes of folks we canvassed. I've been hosted and toasted at parties to celebrate after a successful sign-up had been achieved. In all but a handful of circumstances the jugs, jars and bottles had no brand names, government seals or labels. Those potions were offered as gestures of good will and in appreciation of my work. There is no way in this world that I would even think of betraying the hospitality and sincerity of those people." I was disappointed my neighbour didn't know me well enough to realize what my response would be to his request that I become a snitch, but I guess he was just doing his policeman job, trained to trust no one, not even a friendly neighbour.

Another home-brew story. Shortly after the Yorkton Farm Electrification Branch office opened in 1953, John Chernoff worked for me in a clerical role. He grew up in Veregin, where his father was the Federal Grain elevator agent. I had two wonderful Ukrainian meals in his parent's home in Veregin when I was working on the farm project that brought power to the village. John was my translator during the sign-up and easement clearance phases of that farm project. Before working for SPC John had been a fuel oil truck driver for Northern Petroleum of Kamsack, making deliveries to the surrounding rural community. Apparently the RCMP suspected he was also delivering home-brew, but each time he was stopped and the truck searched the police found not a trace. Ingeniously, the gallon jugs were suspended from the chains that were connected to the fuel compartment lid covers. When the investigating officers took off the lids and peered into the tanks, the jugs were safely hanging on the chains out of sight below the surface of the fuel.

* * *

THE CANADIAN ELECTRICAL Association, an organization of electric power utilities, is a forum for the exchange of information and technology. In 1959 a promotional arm, called the Marketing Services Department was inaugurated to promote increased electrical usage across Canada. One of its activities was the development of high performance standards for a selected number of appliances that would be of better quality material and workmanship than were the regularly produced off-the-shelf products. Electrical manufacturers made these higher quality items with the assurance the utilities across the land would encourage and promote the purchase of them by their consumers. One such appliance was the "Cascade" water heater, which carried a 10-year warranty, a substantially superior product. Over several years hundreds of thousands of "Cascades" were installed in homes, on farms and in commercial premises. Another item was "Sentinel" farmyard lights. Of rugged construction, it had a 175 watt mercury vapour lamp, operated by a photo-electric cell that turned them on at dusk and off at dawn. They provided increased safety and security protection on farms against nighttime vandals and predators.

In the mid-60s SPC's Load Development Department initiated a campaign that connected over 40,000 "Sentinel" units on farms. Promoted in *FL&P*, farmers were given a 10 dollar incentive coupon for every light they purchased from their electricians. One from Willowbrook, whom I had worked with when I was in Farm Electrification Branch office in Yorkton, left his shop Monday mornings with a truck-full of units. He installed one on as many farm yard poles as he had in his truck, telling the farmer he would be back the following week to take it down if it was not wanted. He didn't remove very many as they provided much more light than the incandescent bulbs they replaced.

To put into perspective the success of this one promotional program, the 40,000 "Sentinel" units would consume 32 million kWhs annually and amounted to about 8,000 kW of load on SPC's system, almost four times the 2,100 kW of capacity of the old Yorkton Power Station in 1954 when the Farm Electrification Branch office moved into that building.

* * *

THE JANUARY 1, 2000 issue of *Maclean's* magazine featured a section, "Soaring to Excellence," a series of short stories about Canadians who made a difference in the twentieth century. The late Tommy Douglas was one of those so honoured. His daughter, Shirley, reminisced about him for two achievements.

She first mentioned Saskatchewan's rural electrification program followed, surprisingly, by his more widespread recognized role on the founding father of medicare. About electrification, she said, "People's lives were so hard in Saskatchewan in the Depression. There was no electricity. Everyone said getting electricity was impossible because the province is so enormous. You had all these big farms and so far to go to reach the next one. My father said in the '44 election that he would bring electricity to the province. But no one believed it. Even one of our own supporters turned to my mother and said: 'He has to say that. But we'll just ignore it because we know it can't be done.' But he did electrify the province. And once, when he was flying over Saskatchewan, someone asked him what is the greatest thing you feel you have done for this province? He said: 'Look down. Right there. The twinkling lights.' "

The first week of February was designated as National Electrical Week in the United States where it was extensively promoted by the American electrical industry, but there was only tacit participation in Canada. In 1962 when in the position of Public Relations Superintendent I felt SPC should capitalize during this week to recognize the important conclusion of Saskatchewan's farm electrification program, "Operation Complete Coverage." I placed SPC advertisements in *FL&P* and all the provincial daily and weekly newspapers with commercials aired over the broadcast media which drew attention to this meaningful milestone. As part of this campaign I wrote to every electrified farm customer and requested they leave their yard lights on all night every night that week, as a symbolic gesture to recognize the achievement.

When my 60,000 letters arrived in the mail room I was questioned about the postage, paper and envelope costs for this alleged frivolous notion. The query quickly died when I justified the expense economically. The expected revenue from the yard lights being left on would far exceed the costs of the mail out. I believed the emotional benefits of public recognition of this milestone on its own could have justified the expense. The letters went out.

Farmers responded, pleased to signal this major change in their lives. I heard from many expressing their thanks for the reminder of a service which had by then been taken for granted. One even wrote me that he walked over snowbound roads for over a mile to his unoccupied yet still connected farm so the yard light could be turned on and left to burn all week.

I wrote Trans Canada Airlines and asked to have their cockpit officers announce to passengers on after-dark flights over Saskatchewan to look below at the lights which signalled the transformation from the expanse of darkness that once had stretched endlessly across the plains. The airline agreed and their pilots did so. Those were the "Twinkling Lights" Shirley Douglas's father had talked about.

The Best Part Ends

FROM THE OUTSET of Harry Jessop's arrival in Yorkton I did not take to him so our relationship was, at best, cool. I had the most experience with the branch's various routines so Garnet Parcher had given me free hand to run the office because he had little administrative experience as his prior service time with SPC was as a city operation's crew lineman. Parcher and I trained Jessop on his field duties and the office routines, respectively. When Parcher fell ill in 1954 he appointed me as his substitute, fully expecting to recover. In hindsight Jessop would have been the more logical choice as he was outside the union's scope, in management, but he was a greenhorn still on probation. As Garnet's health waned he did not want to nor did he have the will to reverse his decision. While still a union member, I substituted in the senior field rep position for over a year, so it is little wonder there was ill will between Jessop and me, awkward in an office of nine employees.

Jessop was appointed to replace Garnet Parcher after his death and, therefore, was in charge of the office. Throughout the winter of 1955-56 I organized about two thirds of our area's projects. He arranged my meeting schedules and assigned me jobs in our far north western frontier on the edge of the boreal forest or to the Moose Mountain country in the south, saving those close to Yorkton for himself so he could be conveniently at home with his family every night. For weeks on end I left home Monday morning and returned late on Friday nights or early Saturday morning. Once, he scheduled a Monday morning meeting west of Nipawin, nearly five hours away, requiring my departure on Sunday night. It only happened the once. Jessop's defence of this falderol was the need for him to be in the office as he was in charge, but we both knew, as did the other staff, this was puffery. Harry Monette, who had replaced me as senior farm clerk, was more than able and, in fact, was paid as I had been, to run the office while both Jessop and I were away.

Jessop petulantly took the pleasure out of my work. On the road so much of the time gave me too much of it to ponder my dilemma, and thinking bad thoughts was unappealing. Sadly, the animosity between us became so ingrained and pervading it became routine. I felt like the storied pedestrian who walked on the safety of a green traffic light but was killed by a car running the red. Dead right, but still dead.

That August I had "had enough already!" My 1957 program jobs didn't need me to shepherd them any longer, the easement clearance work was finished and my self-help projects were underway. As well I was at the stage where my work was becoming routine and not the challenge it once was. I called our Regina boss Dan Dojack, and said, "I'm up to here with how I'm being treated, Dan. I don't know if I'm of any use somewhere else in the company, but I've had enough. I want out or I'm leaving." Roy Sarsfield called me only minutes later, said he and Dojack were coming to Yorkton the next day. We met for lunch in the Broadway Cafe and I told them of my dissatisfactions. Sarsfield said, "Dave, please be patient for a short while. I want to keep you in my division." He wasted no time.

In a matter of days Glen Gorham, the Yorkton Area's commercial field rep, was moved to a Regina Head Office job. I was appointed on August 28, 1957 to this non-advertised position and moved to the downtown Yorkton Electrical Distribution Office on Third Avenue, the original location of the Farm Electrification Branch office.

I was grateful to Roy Sarsfield to help me escape and more so for the challenges of the new work and for another small step up my personal ladder.

* * *

I HAD BEEN in the midst of delivering this massive four-year farm electrification program, exceeding, by nearly three per cent, the challenging targets given to SPC as a major campaign promise of the CCF government. The actual achievement province-wide was:

	Number of Individual Farms		Number of Communities	
Year	Target	Connected	Target	Connected
1953	5,500	5,728	31	50
1954	6,500	6,555	25	41
1955	7,500	7,706	20	63
1956	7,500	7,800	20	67
Total	27,000	27,789	96	221

* * *

I HAD HELPED pioneer, develop and implement the Area Coverage program into a viable, smooth-working process. Rural electrification would continue for another three years under this plan: 6,591 farms were served in 1957, 5,191 in 1958 and 3,775 the following year. In 1959, "Operation Complete Coverage" was launched as a final three-year thrust to wind up this massive undertaking. It would blanket the province, one third of the RMs each year. By the end of 1966 more than 66,000 farms had been served. Well over one million poles were needed for the 73,000 miles of single-wire, 13.8 kV rural power lines that connected them. By 1955 all incorporated towns had been served and, by 1957, all villages. In total more than 940 cities, towns, villages, hamlets and other unincorporated settlements were connected to central station power.

As farmers availed themselves to this new energy source and put it to work they helped speed up the need for larger and more efficient generating units in the power stations and a stable, more reliable, high-voltage, transmission grid system which benefited all SPC customers. The use of electricity on farms grew quickly, the increases compounding year after year, a major contributor to the electrical system's whirlwind growth. In 1958, over 500 farms had their transformers increased from 5 to 10 horsepower but only eight years later, in 1966, an impressive 20 per cent of all farms SPC served had transformers of 10 or more horsepower.

Ancillary benefits of farm electrification to the province were the commerce and industry that it sprouted and the employment opportunities they needed to support SPC's expansion program. Pioneer Electric of Regina manufactured thousands of transformers, one for each farmyard. Canada Wire and Cable, in an old Weyburn airport hanger, produced thousands of miles of conductor for the intricate spider-web of lines that connected the farms. IPSCO's Regina steel mill supplied Pioneer's and Canada Wire's steel requirements. Phase converters, to give farm workshops three phase capability for welders and motors from the one-wire single phase farm lines, were made by an Estevan electrical contractor. Many power line contracting firms were started, employing hundreds of workmen, to help SPC crews get the lines and facilities constructed. Distributors and wholesalers of electrical equipment and appliances opened new branch warehouses and competed with the established firms to supply the products needed by hundreds of electricians and retail appliance dealers to meet consumer demands as this new market opened. And then, all these goods needed transporting to their destinations. These are but a smattering of the business spinoffs that farm electrification contributed to the province's growing economy.

Providing the Wheat Province with safe, reliable electrical service was a prodigious accomplishment almost in defiance of economic considerations because SPC has the lowest density of customers of any other North American electric power utility. This miraculous achievement is one in which future generations will not likely ever appreciate the effort involved.

* * *

THE SUN HAD SET on this part of my career. While there was nothing any more remarkable or noteworthy about my work and experiences than happened with any of the other farm field reps in the province, those wonderful days engrossed in pioneering, refining and delivering Saskatchewan's farm electrification program were, without question, the happiest and most satisfying time of my 35-year working career in Canada's energy industry.

I took particular satisfaction in my determination, initiative, optimism and marketing skills to complete a project's sign-up after the farm committee had given up hope and needed me to finish the job, stuff that turned my dials up to 10. Much of this work was done in isolation and gave me huge doses of self confidence, taught me a great deal about corporate governance and the importance of making decisions that were reasoned and fair even though they broke the rules. I had earned the respect of those I worked for and others in more senior managerial positions who supported me as I continued to climb the rungs of my personal ladder in later years: to Assistant to the Sales Manager, Public Relations Superintendent, Assistant to the Electric System Manager, Electric System Business Manager and, in 1980, Corporate Vice-President of Public Affairs. My rise through SPC's ranks from a junior accounting clerk to VP, while not meteoric, was a pretty good trajectory and I could have done worse.

With overflowing pride and satisfaction I will always fondly cherish the looks on the faces of a farmer, his wife and children when electricity was, for the first time, turned on at their farm. Their initial looks of disbelief, then glimmers of excited anticipation, followed by their unbridled pleasure are indelibly etched in my memory. With a flick of a switch they realized that much of their hard, back-breaking, physical labour had been eliminated and an easier and better way of life was then at the control of their fingertips. Knowing my work repeated this euphoria thousands of times, touching the lives of rural people in ways I will never know that forever changed the complexion of the Wheat Province is a priceless personal treasure. I was privileged to have had the opportunity of a lifetime to play a small part in this extraordinary effort that helped Saskatchewan farm families *"To Get The Lights."*

It was an electrifying ride!

Selected Bibliography

Clements, Muriel. *By Their Bootstraps.* A History Of The Credit Union Movement In Saskatchewan. Clarke, Irwin & Company Limited; 1965.

Finch, Claude. *Living Skies and Saskatoon Pies.* Stories and Rhymes of bygone times. Self published; 2000.

Lane, Grace. *Minding Your Own Business.* The Power Credit Union Story. Zephyr Printing, Regina: 1976.

White, Clinton O. *Power For A Province.* A History Of Saskatchewan Power. Canadian Plains Research Centre, University Of Regina: 1976.

Index

Index

Index

Index

Index

Index

ISBN 1-41206371-X